UNSTEADY TRANSONIC FLOW

UNSTEADY TRANSONIC FLOW

MÅRTEN T. LANDAHL

MASSACHUSETTS INSTITUTE OF TECHNOLOGY

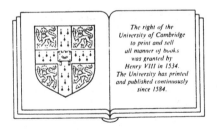

The right of the
University of Cambridge
to print and sell
all manner of books
was granted by
Henry VIII in 1534.
The University has printed
and published continuously
since 1584.

CAMBRIDGE UNIVERSITY PRESS

CAMBRIDGE

NEW YORK   NEW ROCHELLE   MELBOURNE   SYDNEY

CAMBRIDGE UNIVERSITY PRESS
Cambridge, New York, Melbourne, Madrid, Cape Town, Singapore, São Paulo, Delhi

Cambridge University Press
The Edinburgh Building, Cambridge CB2 8RU, UK

Published in the United States of America by Cambridge University Press, New York

www.cambridge.org
Information on this title: www.cambridge.org/9780521356640

First published in hardcover 1961 by Pergamon Press Ltd.
This paperback edition, with preface, first published by Cambridge University Press 1989
Re-issued in this digitally printed version 2009

A catalogue record for this publication is available from the British Library

Library of Congress Cataloguing in Publication data
Landahl, Mårten.
Unsteady transonic flow / Mårten T. Landahl.
p.   cm. – (Cambridge science classics)
Reprint. Originally published: New York: Pergamon Press, 1961.
Bibliography: p.
Includes indexes.
1. Unsteady flow (Aerodynamics)   2. Aerodynamics, Transonic.
I. Title.   II. Series.
TL574.U5L36   1988
629.132′304—dc19   88-21871   CIP

ISBN 978-0-521-35664-0 paperback

CONTENTS

6. THE SEMI-INFINITE RECTANGULAR WING

7. THE RECTANGULAR WING OF ARBITRARY ASPECT RATIO WITH CONTROL SURFACE

8. THE DELTA WING OF ARBITRARY ASPECT RATIO

9. WINGS OF GENERAL PLANFORMS

10. CONTROL SURFACE BUZZ

11. EXPERIMENTAL DETERMINATION OF AIR FORCES ON OSCILLATING WINGS AT TRANSONIC SPEEDS

PREFACE

THE present volume is essentially a collection of papers on the title subject by the author and others during recent years. The interest in unsteady transonic flow arises mainly in connection with calculation of flutter which is more likely to occur in the transonic range than in any other speed range. Most of the work done up to now has been based on linearized theory. Although linearized theory breaks down for steady transonic flow it is known to be valid for unsteady flow provided the unsteadiness is sufficiently large. Unfortunately there is one important type of transonic flutter, namely control surface flutter or "buzz", for which non-linear effects are very important, notably the presence of shock waves. However, recent developments, as those described in Chapter 11, offer some hope for methods of analysis of this phenomenon. Nevertheless, other types of transonic flutter may be adequately treated by use of linearized theory and it is hoped that this monograph will be of some use to aeroelasticians as well as of assistance for further research in this field.

Although the flutter aspect of the problem has been stressed, another related problem, namely that of calculating dynamic stability derivatives (for which linearized theory seems to be valid for wings with highly swept leading edges) is also covered. Other interesting problem areas, like those of transient air forces and forces on accelerating bodies, have been left out completely. Of the content, parts of Chapters 1, 2, 4 and 11 are new and have not been published previously.

I would like to express my sincere thanks to Professors Sune B. Bernt and Holt Ashley for their encouragement and helpful advice during the course of my work on the subject in recent years and also for their careful reading of the part that constituted my doctoral thesis at the Royal Institute of Technology in Stockholm. Also, I would like to express my gratitude to Dr. D. E. Davies of the Royal Aircraft Establishment, Farnborough, for his helpful criticism of the original publications and for going through all my calculations and discovering several algebraic mistakes. Miss Ingrid Jonason typed the manuscript and Miss Elisabeth Redlund and Miss Ulla-Britt Anderson drew all the figures. Thanks are due to them and to all others at the Aeronautical Research Institute of Sweden who helped me with the manuscript in various ways.

Most of the work presented here was done during my stay at the Aeronautical Research Institute of Sweden and supported by the Swedish Air Board through Saab Aircraft Company. The part reported in Chapter 5 was supported by the U.S.A.F. Office of Scientific Research. The support of these organizations is gratefully acknowledged.

PREFACE TO THE PAPERBACK EDITION

This is a reprint of the original 1961 monograph which was based on the doctoral thesis of the author. It deals with the analysis of the unsteady aerodynamic forces acting on oscillating thin two- and three-dimensional wings in transonic flow. Such forces are needed for the prediction of flutter, which tends to occur more readily at transonic speeds than in any other speed regime. The book emphasizes analytical methods for treating the linearized problem for simple planforms such as rectangular and delta wings. It also includes a treatment of slender wing-body combinations.

There is a thorough discussion of the limitations of the validity of linearized acoustical theory showing how this theory becomes valid for large reduced frequencies. At transonic speeds, strong coupling arises between the unsteady flow and the steady flow due to thickness and lift as manifested in the behavior of the slow moving, "receding" wave part of the pressure with a variable wavelength as it propagates through the nonuniform steady flow. This difficulty does not arise for subsonic and supersonic small-perturbation unsteady flows, for which the linear theory of acoustical disturbances propagating through the uniform free stream holds. An important finding in the book is that the contribution of the receding wave part of the pressure to the overall lift and moment becomes negligible at high frequencies, as the positive and negative portions of the receding short-wave pressure contributions then tend to cancel. At transonic speeds, complications also arise from shock waves impinging on the moving wing surface, and the boundary conditions that apply on an unsteadily moving shock wave are considered in the book.

Nowadays the standard engineering practice is to carry out transonic flow calculations directly on the computer using numerical techniques based on finite differences or finite volume and the like. Computer codes for both two- and three-dimensional unsteady transonic flow calculations are available and routinely used in industry for flutter predictions. Analytical treatments of the simplified equations presented in this monograph may nevertheless still be of value because of the physical insight gained through them. Also, the 'exact' solutions for special planforms provided may serve as useful reference solutions for validation of complicated computer codes. Many of the numerical methods developed for steady transonic flow also start from the formulation for an unsteady flow and seek the steady flow as the limit as $t \rightarrow$-, so that an understanding of transonic unsteady flow phenomena is of value for the steady flow problem as well.

A large number of papers on unsteady transonic flow have of course appeared in the literature since this monograph was originally published. For example, the comprehensive review by Borland (USAF AFFDL-TR-78-189, Vol. 1, Fluid Dynamics Lab., Wright-Patterson AFB, Ohio, 1979) lists more than 200 articles published between 1970 and 1979. Tijdeman and Seebass (*Ann. Rev. Fluid Mech.*, **12**, 1980) reviews the two-dimensional problem and also discusses the effects of viscosity. The author's own review article in Symposium Transsonicum II (ed. K. Oswatitsch, Springer Verlag, 1976) pays special attention to the propagation of acoustical waves in a nonuniform transonic flow and their importance for the related problem of the stability of a smoothly decelerating shock-free flow, a problem originally treated by Spee (NLR TR 69122 U, Amsterdam 1969). Many examples of unsteady transonic flow calculations for both two- and three-dimensional wings can be found in the aeronautical literature, for example in the AIAA Journal and the AIAA Journal of Aircraft.

Mårten T. Landahl
1988

INTRODUCTION

IN RECENT years a large amount of work has been expended on the study of aerodynamic forces on oscillating wings. These forces are needed for the investigation of the dynamic or aeroelastic stability of an airplane. Such problems have received increased attention with the advent of transonic and supersonic airplanes.

There are two reasons why the phenomenon of flutter has become more important as airplanes have surpassed sonic speed. Firstly, of course, the more slender shapes required for supersonic flight have made the airplanes more flexible and therefore more prone to aeroelastic instabilities. Secondly, the aerodynamic forces at transonic speeds are such as to favor the occurrence of flutter, so that the critical speed usually has a minimum at or near $M = 1$. For an excellent discussion of transonic flutter problems the reader is referred to a paper by Garrick (Ref. 18).

The affinity to flutter in the transonic speed range may be explained from well-known aerodynamic properties of transonic flow. The lifting pressures due to a given amount of deflection are known to be at maximum at or near $M = 1$ (cf. the lift curve slope). This must lower the flutter velocity since an increase of all aerodynamic derivatives by the same amount has the same effect as, for example, increasing the air density. However, an effect which is probably even more important is that due to the large phase lags between motion and unsteady air pressures that occur at transonic speeds. When an object travels at a speed near that of sound the flow perturbations created move forward at about the same speed as the object itself. Hence there will be a slow accumulation of disturbances and, if the flow is given sufficient time to build up, the well-known typical transonic non-linearities will occur. Since a pressure wave set up at a point will spend a long time before it travels off the object, it is evident that large and possibly destabilizing phase differences between motion and pressure can easily be created. These are directly responsible for one-degree-of-freedom flutter of control surfaces (control surface "buzz") and also for the low or negative damping in pitch sometimes encountered by tailless aircraft of high or medium aspect ratios.

Because the computation of aerodynamic forces on oscillating three-dimensional wings is so complicated, even on the basis of linearized theory, most flutter calculations in industry today are made by use of aerodynamic derivatives obtained from two-dimensional (strip theory) analysis. For high supersonic Mach numbers or for large-aspect-ratio wings in subsonic flow this procedure may be justified. At transonic speeds, however, cross-flow effects are always very large as is well known in the case of steady flow. Therefore, the use of strip theory can lead to large errors in the computed flutter speed near $M = 1$. For example, a strip-theory flutter calculation of a configuration involving a control surface will always show one-degree-of-freedom flutter of the control surface at transonic speeds unless the hinge stiffness is very high or artificial damping is provided. As shown in Chapter 7, however, the three-dimensional theory, on the contrary, gives positive hinge moment damping at $M = 1$ for rectangular control surfaces of aspect ratio less than 3.5.

Since there is no prospect of integrating the full non-linear transonic equations of fluid motion, any three-dimensional lifting-surface theory would have to be based on the linearized equations. For sub- or super-sonic flow linearized theory is known to hold well for thin wings. For transonic flow, however, the above-mentioned non-linear accumulation of disturbances precludes the use of linearized theory in the steady, non-lifting case no matter how thin the wing is. In the oscillating wing case the situation is somewhat better, though. Firstly, one is concerned with the lifting part of the flow. According to the transonic equivalence law, Ref. 54, linearized theory is capable of describing the steady lifting flow for wings of low aspect ratio, and wind tunnel experiments do, indeed, confirm fairly well the predictions of the theory at least for wings with swept leading edges. Secondly, if the wing oscillates rapidly the non-linear disturbance accumulation will not have time to develop and hence the linearized equations will be applicable. The conditions necessary for this to apply are discussed in Chapter 1.

The main part of the present monograph is devoted to the study of lifting surface theories. Most of this is based on recent theoretical work by the writer (Refs. 30–39) but available investigations by other workers in the field have also been included for completeness. The writer's methods have been developed with the aim of covering the reduced frequencies of interest in flutter research. For the sake of simplicity, however, most numerical results given are those for stability derivatives, i.e. for rigid-body wing motions, but evaluated at frequencies of interest in flutter work. One exception to this is Chapter 5 on wing-body interference at sonic speed in which only results for stability derivatives at low reduced frequencies are given.

No comprehensive treatment of the present subject would be complete without reference to experimental results. In Chapter 11 material available in the open literature is collected on the subject for cases where direct comparisons with theory are possible.

The writer realizes that there are transonic unsteady-flow problems that merit attention other than the oscillating wing problem and which have been left out in this monograph. In principle, the oscillating case can be considered as the Fourier transform with respect to time of an arbitrary time dependent motion so that it is possible to make use of the given results for any type of motion. However, such a method is seldom practical. References dealing with transient phenomena at transonic speeds can be found, for example, in Miles' recent book (Ref. 49).

CHAPTER 1

THE EQUATIONS OF MOTION AND THEIR LINEARIZATION

1.1. Introduction

THE first systematic investigation of the conditions under which the equations of motion for two-dimensional unsteady transonic flow can be linearized was made by Lin *et al.* (Ref. 41). Their analysis was later extended to three-dimensional flow by Miles (Ref. 47) and Mollö-Christensen (Ref. 50). These investigations were all based on the assumption that terms in the differential equation that are definitely small compared to the others can be neglected.

The only rigorous method to ascertain the validity of linearization is of course to start from the exact solution of the non-linear problem (assumed unique) and then investigate whether the solution could be expanded so as to give the linearized solution as the initial term. The conditions necessary for linearization will then automatically follow from the higher-order terms in the expansion. Such an approach was actually tried in Ref. 40 but the method used in obtaining the series expansion (iteration) implicitly assumed that the solution could be linearized for some combinations of the parameters involved, and that the iteration converged in some manner.

The considerations given below (Ref. 35) are based on physical rather than mathematical arguments. It is not claimed to be more rigorous than the earlier attempts but it is believed to give a better physical insight into the problem. Also it affords a very simple unified expression of the requirement sufficient for linearization, namely Eq. (1.40).

1.2. Equations of motion

The basis for deducing the perturbation potential equation will be the equation of continuity:

$$\rho_t + (\rho U)_x + (\rho V)_y + (\rho W)_z = 0 \qquad (1.1)$$

where U, V and W denote the velocity components in a Cartesian co-ordinate system x, y, z.

Let the flow consist of a main stream directed along the positive x-axis. The flow is slightly perturbed by a thin body which is mainly oriented along the x-axis and executes small transverse unsteady motions. The velocity of the unperturbed free stream, U_0, is set equal to unity and so is the characteristic length, b, of the body. Thus the velocity of sound in the free stream, c_0, is then equal to $1/M$, where M is the free stream Mach number. No loss of generality will be incurred by this choice of non-dimensional co-ordinates, provided t is interpreted as a non-dimensional time obtained by multiplying the physical time by U_0/b.

Assuming the perturbation of the flow to be everywhere small it is permissible, within a certain order of approximation, to assume the existence of a perturbation velocity potential $U_0 b \phi(x, y, z, t)$ such that

$$U = U_0(1 + \phi_x); \qquad V = {}^{\prime\prime} \phi_y; \qquad W = U_0 \phi_z \qquad (1.2)$$

Thus the continuity equation becomes

$$\rho_t + [(1 + \phi_x)\rho]_x + (\rho\phi_y)_y + (\rho\phi_z)_z = 0 \qquad (1.3)$$

In order to eliminate the ρ-derivatives we use the Bernoulli equation

$$\phi_t + \phi_x + \tfrac{1}{2}(\phi_x^2 + \phi_y^2 + \phi_z^2) + \frac{c^2}{\gamma - 1} = \frac{1}{M^2(\gamma - 1)} \qquad (1.4)$$

In isentropic flow

$$d\left(\frac{c^2}{\gamma - 1}\right) = c^2 \frac{d\rho}{\rho} \qquad (1.5)$$

Hence, by introducing Eqs. (1.5) and (1.4) in Eq. (1.3) we arrive at the following partial differential equation for ϕ:

$$(1 - M^2)\phi_{xx} + \phi_{yy} + \phi_{zz} - 2M^2\phi_{xt} - M^2\phi_{tt}$$
$$= M^2[\tfrac{1}{2}(\gamma - 1)(2\phi_x + 2\phi_t + \phi_x^2 + \phi_y^2 + \phi_z^2)(\phi_{xx} + \phi_{yy} + \phi_{zz})$$
$$+ (2\phi_x + \phi_x^2)\phi_{xx} + \phi_y^2\phi_{yy} + 2\phi_y\phi_z\phi_{yz} + \phi_z^2\phi_{zz}$$
$$+ 2(1 + \phi_x)(\phi_y\phi_{yx} + \phi_z\phi_{zx}) + 2(\phi_x\phi_{xt} + \phi_y\phi_{yt} + \phi_z\phi_{zt})] \quad (1.6)$$

Equation (1.6) is strictly valid only for isentropic flow but is a good approximation even for flows with weak shocks, although it would then be inconsistent to retain all of the higher order terms since the vorticity induced by the shocks would give contributions of the same order as those of some higher order terms. An analysis of this in Ref. 7 for a slender body of revolution at $M = 1$ shows that the flow is, indeed, isentropic to at least second order in perturbation velocities.

When trying to simplify the formidable Eq. (1.6) the commonly used philosophy is to assume the perturbations to be so small that the quadratic and cubic terms on the right-hand side may be neglected in comparison with the linear terms. Hence one obtains the well-known differential equation of acoustical theory

$$(1 - M^2)\phi_{xx} + \phi_{yy} + \phi_{zz} - 2M^2\phi_{xt} - M^2\phi_{tt} = 0 \qquad (1.7)$$

The assumption of small perturbations leads to considerable simplifications but it does not justify the neglect of all non-linear terms in the transonic case as will be discussed subsequently.

1.3. Deduction of the small perturbation equation

The differential equation for small perturbations is the one which gives the correct first-order solution as the perturbation velocities become vanishingly small. From a physical viewpoint it seems plausible that this equation is obtained by retaining in each individual term in the continuity equation only the lowest order terms as the perturbation velocities tend to zero.

Consider first the term $(\rho\phi_y)_y$ in Eq. (1.3). By aid of Eq. (1.5) it may be written

$$(\rho\phi_y)_y = \frac{\rho}{c^2}\left[c^2\phi_{yy} + \frac{\phi_y}{\gamma-1}\frac{\partial}{\partial y}(c^2 - M^{-2})\right] \qquad (1.8)$$

or

$$(\rho\phi_y)_y = \frac{\rho}{c^2}\left\{M^{-2}\frac{\partial}{\partial y}\left[\phi_y\left[1 + \frac{M^2}{\gamma-1}(c^2 - M^{-2})\right]\right] + \frac{\gamma-2}{\gamma-1}(c^2 - M^{-2})\phi_{yy}\right\} \qquad (1.9)$$

As the perturbation velocities tend to zero, $c^2 \to M^{-2}$, and hence the first term in Eq. (1.9) tends to

$$M^{-2}\frac{\partial}{\partial y}(\phi_y \cdot 1) = M^{-2}\phi_{yy} \qquad (1.10)$$

where we have assumed that $\partial/\partial y(c^2)$ vanishes along with the perturbation velocities.

The second term in Eq. (1.9) will hence become vanishingly small compared to the first one so that, to the first order,

$$(\rho\phi_y)_y = \frac{\rho}{c^2}(M^{-2}\phi_{yy}) \qquad (1.11)$$

Comparing Eq. (1.11) to Eq. (1.8), expanded by aid of Eqs. (1.5) and (1.4), shows that non-linear terms like $\phi_y\phi_{yt}$, $\phi_y\phi_{yx}$, $\phi_y\phi_x\phi_{xy}$, $\phi_y^2\phi_{yy}$ and $\phi_y\phi_z\phi_{yz}$ may, indeed, be neglected compared to $M^{-2}\phi_{yy}$.

In the same manner we can show that the first-order result for $(\rho\phi_z)_z$ is

$$(\rho\phi_z)_z = \frac{\rho}{c^2}(M^{-2}\phi_{zz}) \qquad (1.12)$$

and that terms like $\phi_z\phi_{zt}$, $\phi_z\phi_{xz}$, $\phi_z\phi_x\phi_{xz}$ and $\phi_z^2\phi_{zz}$ are of higher order than $M^{-2}\phi_{zz}$.

We turn now to the term ρ_t. Using Eqs. (1.4) and (1.5) we obtain

$$\rho_t = \frac{\rho}{c^2}(\phi_{tt} + \phi_{xt} + \phi_x\phi_{xt} + \phi_y\phi_{yt} + \phi_z\phi_{zt}) \qquad (1.13)$$

The terms $\phi_y\phi_{yt}$ and $\phi_z\phi_{zt}$, however, were just shown to be of higher order than $M^{-2}\phi_{yy}$ and $M^{-2}\phi_{zz}$, respectively, and should therefore be similarly neglected.

Also $\phi_x\phi_{xt}$ must be small compared to ϕ_{xt} so that, to the first order,

$$\rho_t = -\frac{\rho}{c^2}(\phi_{tt} + \phi_{xt}) \tag{1.14}$$

The term $[(1 + \phi_x)\rho]_x$ finally, reads, using Eqs. (1.4) and (1.5) and retaining all terms,

$$[(1 + \phi_x)\rho]_x = \frac{\rho}{c^2}[(c^2 - 1 - \phi_x)\phi_{xx} - (1 + \phi_x)(\phi_{xt} + \phi_x\phi_{xx} + \phi_y\phi_{yx} + \phi_z\phi_{zx})] \tag{1.15}$$

In the next last bracket ϕ_x may be neglected, but otherwise the remaining terms may all be of the same order. However, the terms $\phi_y\phi_{yx}$ and $\phi_z\phi_{zx}$ were just shown to be small in comparison with the terms $M^{-2}\phi_{yy}$ and $M^{-2}\phi_{zz}$, respectively, and are therefore neglected. Hence

$$[(1 + \phi_x)\rho]_x = \frac{\rho}{c^2}[(c^2 - 1 - 2\phi_x)\phi_{xx} - \phi_{xt}] \tag{1.16}$$

Note that in this term the bracket may not be approximated by its linearized value since in a transonic theory we must allow c^2 to become arbitrarily close to unity as ϕ_x vanishes. Thus, for c^2 the second-order expression must be used which, according to Eq. (1.4), reads

$$c^2 = M^{-2} - \frac{\gamma - 1}{2}(2\phi_x + 2\phi_t + \phi_y{}^2 + \phi_z{}^2) \tag{1.17}$$

The only simplification possible for this was to neglect $\phi_x{}^2$ compared to $2\phi_x$ since we do not know the relative order of magnitude of the different velocity components. Thus

$$[(1 + \phi_x)\rho]_x$$
$$= \frac{\rho}{c^2}\left\{\left[M^{-2} - 1 - (\gamma + 1)\phi_x - \frac{\gamma - 1}{2}(2\phi_t + \phi_y{}^2 + \phi_z{}^2)\right]\phi_{xx} - \phi_{xt}\right\} \tag{1.18}$$

Now the term $\phi_t\phi_{xx}$ must be of the same order as $\phi_x\phi_{xt}$ and could thus be similarly neglected. This can be made plausible, for example, by introducing similarity parameters as in Section 1.4. The terms $\phi_y{}^2\phi_{xx}$ and $\phi_z{}^2\phi_{xx}$ could be neglected in the same way since they are of the same order as $\phi_x\phi_y\phi_{yx}$ and $\phi_x\phi_z\phi_{zx}$, respectively.

Further simplifications are not possible at the present stage so that the final result for the small perturbation equation is

$$[1 - M^2 - M^2(\gamma + 1)\phi_x]\phi_{xx} + \phi_{yy} + \phi_{zz} - 2M^2\phi_{xt} - M^2\phi_{tt} = 0 \tag{1.19}$$

We recognize the non-linear term which is usually retained for steady transonic flow. Although its coefficient may be given in several theoretically equivalent forms for M close to one

$$\gamma + 1, 2M^2\left[1 + \left(\frac{\gamma - 1}{2}\right)M^2\right], M^2(\gamma + 1),$$

the one arrived at here was found empirically (Ref. 62) to give results in closest agreement with more exact theories and experiments.

The first-order boundary conditions belonging to Eq. (1.19) are easier to derive since they involve only first-order derivatives. Let $B(x, y, z, t) = 0$ define the position of the body surface at any instant. Then, in order that the resultant velocity vector should be tangential to the surface,

$$B_t + (1 + \phi_x)B_x + \phi_y B_y + \phi_z B_z = 0 \quad \text{on } B = 0 \tag{1.20}$$

To the first order, for a general thin body, this equation becomes

$$B_t + B_x + \phi_y B_y + \phi_z B_z = 0 \tag{1.21}$$

For a planar body, for which all points lie close to $z = 0$, the tangential condition, Eq. (1.21), can be further simplified. Let $B = z - H(x, y, z, t)$ so that H and its derivatives are small compared to unity everywhere. To the first order, this gives for Eq. (1.21)

$$\phi_z = H_x + H_t \quad \text{on } z = 0 \tag{1.22}$$

Other boundary conditions necessary for the linearized problem, such as the finiteness conditions for ϕ at infinity, are formulated in Section 1.9. Further boundary conditions for the non-linear problem are considered in Chapter 10.

1.4. Introduction of dimensionless parameters

Let the members of a class of affine wings or wing–body combinations be characterized by the thickness ratio ε and the semi-span-to-chord ratio σ. The transverse unsteady motion of the body is given the dimensionless amplitude δ. In order for the flow perturbations to be small it is necessary for both ε and δ to be small compared to unity. If also σ is small the configuration is said to be slender. For a body of revolution $\sigma = \varepsilon$. As a measure of the rapidity with which the unsteady motion takes place we introduce the parameter k. For oscillatory motion k is equal to the reduced frequency $\omega b/U_0$; for transient type of motion k is the inverse of the dimensionless time required to complete the motion. With the choice of non-dimensional variables made in Section 1.2, $\omega = k$ for harmonic oscillations.

These parameters may be incorporated in the tangency condition, Eq. (1.21). Let H be defined by

$$H = \varepsilon h_1(x, \bar{y}) \, \text{sgn} \, z + \delta h_2(x, \bar{y}, \bar{t}) \tag{1.23}$$

where

$$\bar{y} = y/\sigma; \qquad \bar{t} = kt \tag{1.24}$$

The distribution of the unsteady displacement of the wing mean surface from $z = 0$ is given by h_2 and its associated non-dimensional amplitude by δ. h_1 represents the thickness distribution. Introducing Eq. (1.23) in Eq. (1.21) gives

$$\phi_z = \varepsilon f_{1x} \operatorname{sgn} z + \delta(f_{2x} + k f_{2\bar{t}}) + \sigma^{-1}\phi_y(\varepsilon f_{1\bar{y}} \operatorname{sgn} z + \delta f_{2\bar{y}}) \qquad (1.25)$$

The planar case occurs when $\sigma \gg \varepsilon$ (and $\sigma \gg \delta$). Then

$$\phi_z = \varepsilon f_{1x} \operatorname{sgn} z + \delta(f_{2x} + k f_{2\bar{t}}) \qquad (1.26)$$

From the form of Eq. (1.26) it seems natural to separate ϕ into two parts, ϕ_1 and ϕ_2, where ϕ_1 corresponds to the mean steady part symmetric in z and ϕ_2 takes care of the non-steady effects. Hence,

$$\phi_{1z} = \varepsilon f_{1x} \operatorname{sgn} z \qquad (1.27)$$

$$\phi_{2z} = \delta(f_{2x} + k f_{2\bar{t}}) \qquad (1.28)$$

For the non-planar case the steady and non-steady case could still be separated provided $f_{2\bar{y}}$ is not large. Then

$$\phi_{1z} - (\varepsilon/\sigma)f_{1\bar{y}}\phi_{1y} \operatorname{sgn} z = \varepsilon f_{1x} \operatorname{sgn} z \qquad (1.29)$$

$$\phi_{2z} - (\varepsilon/\sigma)f_{1\bar{y}}\phi_{2y} \operatorname{sgn} z = \delta(f_{2x} + k f_{2\bar{t}}) \qquad (1.30)$$

These equations could also be written

$$\phi_{1n} = f_{1x} \sin \nu \qquad (1.31)$$

$$\phi_{2n} = \delta(f_{2x} + k f_{2\bar{t}}) \sin \nu \qquad (1.32)$$

where ϕ_n is the velocity component normal to the curve $B = 0$; $x = $ constant making the angle ν with the y-axis as shown in Fig. 1.1.

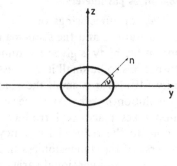

Fig. 1.1.

The boundary conditions for the transverse unsteady flow represented by ϕ_2 become fully linearized if δ is assumed so small that Eq. (1.32) can be satisfied on $z = \varepsilon f_1 \operatorname{sgn} z$ instead of on $B = 0$. In particular, this is a valid approximation if we assume $\delta \ll \varepsilon$. Such an assumption is reasonable since in most unsteady flow problems the stability of infinitesimal motions is studied. We are then led to the possibility of a complete linearization of the unsteady problem by letting ϕ_2 be a small perturbation on ϕ_1. This gives the following linear differential equation for ϕ_2

$$(1 - M^2)\phi_{2xx} - M^2(\gamma + 1)\frac{\partial}{\partial x}(\phi_{1x}\phi_{2x}) + \phi_{2yy} + \phi_{2zz} - 2M^2\phi_{2xt} - M^2\phi_{2tt} = 0$$

$$(1.33)$$

which is the most general linearized equation for a small perturbation unsteady lifting flow. In the following we will investigate the further possibilities of simplifying Eq. (1.33) for particular relations between the parameters introduced.

1.5. Linearized equation for the lifting unsteady flow

The non-constant coefficient term in Eq. (1.33) makes this still too complicated for practical use. By putting certain restrictions on the range of parameters, however, it may be further simplified. Since the non-linearity arises from the term $[(1 + \phi_x)\rho]_x$ in the continuity equation (1.3) this term may be studied separately and the same technique used as in Section 1.3. Hence, from Eq. (1.18) it follows that the first-order contribution to $[(1 + \phi_x)\rho]_x$ from the lifting flow may be written (neglecting the quadratic term in ϕ_2 in accordance with Eq. (1.33))

$$\Delta[(1 + \phi_x)\rho]_x = \frac{\rho}{c^2}\frac{\partial}{\partial x}\{[M^{-2} - 1 - (\gamma + 1)\phi_{1x}]\phi_{2x} - \phi_{2t}\} \quad (1.34)$$

In transonic flow $1 - M = 0(\phi_{1x})$; thus it suffices to study the case $M = 1$.

We may assume $\phi_{2t} = 0(k\phi_{2x})$. Consequently Eq. (1.34) will be dominated by the unsteady linear term and hence the right-hand side approximated by $-\rho c^{-2}\phi_{2xt}$, whenever

$$k \gg |\phi_{1x}| \quad (1.35)$$

everywhere in the flow.

The ensuing differential equation for ϕ_2 in the transonic region will then be

$$\phi_{2yy} + \phi_{2zz} - 2M^2\phi_{2xt} - M^2\phi_{2tt} = 0 \quad (1.36)$$

For two-dimensional flow $\phi_{1x} = 0(\varepsilon^{2/3})$, and the requirement (1.35) becomes

$$k \gg \varepsilon^{2/3} \quad (1.37)$$

From the transonic equivalence law (Ref. 54) one finds that in the neighborhood of a slender wing $\phi_{1x} = 0(\sigma\varepsilon \ln \sigma\varepsilon^{1/3})$, (Ref. 30), so that linearization then requires

$$k \gg \sigma\varepsilon \ln \sigma^{-1}\varepsilon^{-1/3} \quad (1.38)$$

The requirement (1.37) was obtained originally in Ref. 41 in a different manner. Both (1.37) and (1.38) were found in Ref. 40 using a considerably more complicated method involving the expansion, by means of iteration, of the full solution of Eq. (1.33).

Although (1.35) is a sufficient condition for linearization it is not always necessary. For the lifting flow around a pointed slender wing, the expansion method used in Ref. 40 indicated that the slender-wing theory could be used even in the steady or quasi-steady flow, i.e. when $k \ll \sigma\varepsilon \ln \sigma^{-1}\varepsilon^{-1/3}$. The analysis which was based on Eq. (1.19) showed that the relative error of slender-wing theory in this case was of the order $\sigma^3\varepsilon \ln^2 \sigma\varepsilon^{1/3} + k\sigma^2 \ln \sigma\varepsilon^{1/3}$. (Actually, the error that stems from neglecting the term $2M^2(\phi_y\phi_{yx} + \phi_z\phi_{zx})$ is of the order $\sigma\varepsilon \ln \sigma\varepsilon^{1/3}$ which might be larger than $\sigma^2\varepsilon \ln^2 \sigma\varepsilon^{1/3}$ originating from $M^2(\gamma + 1)\phi_x\phi_{xx}$.) For the case where $k \gg \sigma\varepsilon \ln \sigma^{-1}\varepsilon^{-1/3}$ but k still is small, a relative error of order $\sigma^3\varepsilon \ln \sigma\sqrt{k} + k\sigma^2 \ln \sigma\sqrt{k}$ was found. Thus, since Eq. (1.36) becomes the Laplace equation for the cross-flow as $k \to 0$, it would be valid for the flow around a slender wing with no restriction on the lower limit of the frequency. However, this is true only as regards the calculation of aerodynamic stiffness derivatives for pointed wings like the delta. (For rectangular wings slender-body theory predicts that the center of pressure is located at the leading edge which is obviously incorrect.) For damping derivatives one needs to know correctly the out-of-phase part of ϕ_2 proportional to k as $k \to 0$ and this cannot be found from linearized theory. This is also evident from the fact that linearized theory gives damping derivatives which become logarithmically infinite for finite values of σ as $k \to 0$ (see Chapter 3).

Note that, since in transonic flow $1 - M = 0(\phi_{1x})$, Eq. (1.35) automatically implies that

$$k \gg |1 - M| \tag{1.39}$$

Thus a sufficient condition for linearization in the transonic region is that

$$k \gg |1 - M_L| \tag{1.40}$$

is fulfilled everywhere in the flow, where M_L is the local Mach number.

Equation (1.7) is a consistent approximation only outside the transonic region, i.e. where

$$|1 - M| \gg |\phi_{1x}| \tag{1.41}$$

Hence it should be used only when

$$|1 - M| \gg \varepsilon^{2/3} \tag{1.42}$$

in the two-dimensional case and when

$$|1 - M| \gg \sigma\varepsilon \ln \sigma^{-1}\varepsilon^{-1/3} \tag{1.43}$$

in the slender-wing case.

1.6. Discussion of the properties of the transonic solution

First let us study the familiar solution of Eq. (1.7) for one-dimensional flow. Assuming the acoustical disturbance to originate at $x = 0$ the solution for subsonic flow reads

$$\phi_2 = f\left(t - \frac{Mx}{1 + M}\right) \quad \text{for } x > 0$$

$$\phi_2 = g\left(t + \frac{Mx}{1 - M}\right) \quad \text{for } x < 0$$

(1.44)

where f and g may be related but otherwise arbitrary functions of time.
Similarly, for $M > 1$

$$\phi_2 = f\left(t - \frac{Mx}{1 + M}\right) + g\left(t - \frac{Mx}{M - 1}\right) \quad \text{for } x > 0$$

$$\phi_2 = 0 \quad \text{for } x < 0$$

(1.45)

The part of the solution given by f represents a wave that is swept downstream with the speed $c + 1 = M^{-1} + 1$ ("advancing wave"). This wave changes only slowly as the speed passes from sub- to super-sonic. The other part of the solution given by g represents a "receding wave" (adopting the nomenclature of Ref. 44) which travels with the velocity $c - 1 = M^{-1} - 1$ either up- or down-stream depending on whether the flow is sub- or super-sonic. Its wavelength vanishes as $M \to 1$.

Consider now the solution of Eq. (1.36) for one-dimensional flow

$$\phi_2 = f\left(t - \frac{x}{2}\right) \quad \text{for } x > 0$$

$$\phi_2 = 0 \quad \text{for } x < 0$$

(1.46)

In this case the receding wave remains at $x = 0$.

In two- and three-dimensional flow these two wave-fronts similarly occur (although they are actually different parts of the same cylindrical or spherical wave). We will assume that the receding wave has a wavelength which is small compared to some typical reference length of the wing, in accordance with (1.39). Then, provided the strength of the disturbance sources varies smoothly in the x-direction, neighboring receding waves will interact and cancel out as $k/|1 - M| \to \infty$ so that the solution will approach the one corresponding to Eq. (1.46). At leading and trailing edges of an airfoil, however, there will occur discontinuities in the strength of the disturbances and cancellation does not occur. This explains the discontinuous behavior in the linearized pressure distributions on an oscillating two-dimensional airfoil when $M = 1$ is passed, as pointed out by Jordan (Ref. 24) and which is further discussed in Chapter 2.

The behavior is illustrated in Fig. 1.2. The calculation is based on the formulas given in Chapter 2 and is valid for translational oscillations with $k = 0.50$ at $M = 0.975$, $M = 1.0$, and $M = 1.025$, respectively. In the supersonic pressure distribution there will occur a short-wave part originating at the leading edge (corresponding to the second part of Eq. (1.45)). As $M \to 1$ the amplitude of

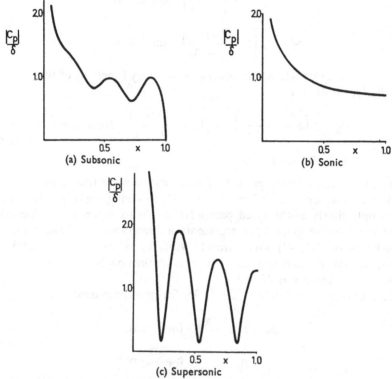

FIG. 1.2. Typical lifting pressure amplitude distributions on an oscillating airfoil obtained from acoustical theory.

the receding wave part, similarly as in Eq. (1.45), remains finite but the wavelength vanishes like $(M - 1)/k$. This behavior led Jordan to discard the acoustical solution for low supersonic Mach numbers in favor of the sonic one since ϕ_{2xx} will become infinite and thereby violate the assumptions of linearized theory. We will show subsequently that the acoustical theory does indeed lead to physically inadmissible solutions in the transonic range and that the transonic Eq. (1.36), besides being the consistent linearized equation for transonic flow according to Section 1.5, does give physically correct solutions. The contribution from the short wave to total forces and moments will be zero as the wavelength vanishes but for small but non-zero values of $(M - 1)/k$ small oscillations will remain (see Section 1.8).

In the subsonic pressure distribution there is a similar part originating at the trailing edge but in this case its amplitude, as well, will vanish when $M \rightarrow 1$ (except at the trailing edge due to the Kutta condition, see Chapter 2). This distribution thus converges to the sonic one (but non-uniformly at the trailing edge) in contrast to the supersonic one.

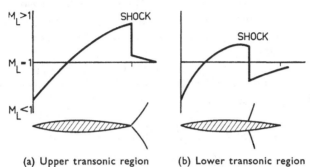

(a) Upper transonic region   (b) Lower transonic region

FIG. 1.3. Typical local number Mach distributions on a non-lifting airfoil in steady flow.

Consider now what will happen in a non-linear transonic flow. The Mach number distribution at $M \simeq 1$ set up by the non-lifting flow may look as in Fig. 1.3a.

This distribution is constant in a small region around $M = 1$ (Mach number freeze). Receding waves will in the real flow move with a speed of $c - U$ (c and U being local values) which both in sign and magnitude may differ from the linearized value $M^{-1} - 1$. The advancing waves, however, will always have a speed close to the linearized value $M^{-1} + 1$. From Fig. 1.3a it then follows, that the receding wave due to the strong disturbance from the leading edge will move upstream, away from the wing. Neither will the trailing edge cause any short-wave disturbances on the wing since the flow is supersonic there. Consequently, the transonic equation (1.36) will give a pressure distribution more in accordance with physical reality than will the acoustical equation (1.7).

A somewhat different situation arises for an oscillating aileron. Here the flow over the aileron will be supersonic according to Fig. 1.3a and one would hence expect some waviness in the curves of forces and moments vs. k of the kind displayed in Figs. 1.6 and 1.7 of Section 1.8. However, this waviness will have a different wavelength than that given by the free stream Mach number. If Eq. (1.35) is fulfilled, where now k is based on the aileron chord, forces and moments will be correctly given by the transonic theory.

In the lower transonic region one encounters a Mach number distribution for the non-lifting flow like the one shown in Fig. 1.3b. Then, receding waves from the trailing edge region travel upstream to the shock wave and get reflected there. For a wing with aileron this interaction of primary and reflected waves may give the aileron a negative aerodynamic damping. This is the phenomenon involved in so called aileron buzz which is further considered in Chapter 10. Naturally, neither Eq. (1.7) or (1.36) would be capable to describe such a flow. In general, the calculation of aileron derivatives on the basis of linearized theory is questionable in the transonic region due to the low reduced frequencies, based on aileron chord, ordinarily being of interest for flutter calculations.

We have thus seen that influence regions and phase angles of the receding wave are incorrectly given by the acoustical theory. Let us now investigate the receding wave amplitude.

Consider, for example, a subsonic three-dimensional, harmonically oscillating, acoustical source of intensity $q(t) = \text{Re}\,\{\bar{q}e^{ikt}\}$ placed at the origin. Its potential is

$$\phi_2 = \text{Re}\left\{\frac{\bar{q}}{4\pi\tilde{R}} \exp\left[ik\left(t + \frac{M^2 x}{1 - M^2} - \frac{M\tilde{R}}{1 - M^2}\right)\right]\right\} \qquad (1.47)$$

where

$$\tilde{R} = \sqrt{[x^2 + (1 - M^2)(y^2 + z^2)]}$$

On the x-axis ϕ_{2x} thus becomes

$$\phi_{2x} = \text{Re}\left\{\frac{\bar{q}}{4\pi x}\left(\frac{1}{|x|} + \frac{ikM}{M \pm 1}\right) \exp\left[ik\left(t - \frac{Mx}{M \pm 1}\right)\right]\right\} \qquad (1.48)$$

where the upper sign is to be used for $x > 0$ (i.e. for the advancing wave) and the lower sign for $x < 0$ (for the receding wave).

In Fig. 1.4 is plotted $\phi_{2x}/\bar{q}$ vs. x ($\bar{q}$ considered purely real) at the instants $kt = 2\pi n$ (n an integer) for $k = 1$ and some different Mach numbers. It is seen that, even if the immediate region close to $x = 0$ is excluded (where any linearized theory must always break down), ϕ_{2x} in the receding wave will tend to infinity as $M \to 1$ or $k/(1 - M) \to \infty$, even for a very weak source, which is incompatible with the assumptions of linearized theory. For the advancing wave, however, ϕ_{2x} stays small and changes only slowly with Mach number.

Conceivably, even small non-uniformities in the main steady flow must have a large effect on the receding wave. To study this, consider the one-dimensional case for which the linearized equation (1.33) for $\phi_2 = \text{Re}\,\{\varphi e^{ikt}\}$ reads

$$\frac{\partial}{\partial x}\{[1 - M^2 - M^2(\gamma + 1)\phi_{1x}]\varphi_x\} - 2ikM^2\varphi_x + k^2 M^2\varphi = 0 \quad (1.49)$$

Since φ_x/φ must be of order (wavelength)$^{-1}$ we may neglect the last term for the receding wave solution. Equation (1.49) then becomes a first-order differential equation for φ_x with the solution

$$\varphi_x(x) = \frac{\text{const.}}{F(x)} \exp\left[2ikM^2 \int^x \frac{d\xi}{F(\xi)}\right] \qquad (1.50)$$

where

$$F(x) = 1 - M^2 - M^2(\gamma + 1)\phi_{1x}$$

Or, since in the transonic region

$$F(x) \approx M^2(c_1^2 - U_1^2) \approx 2M^2(c_1 - U_1) \qquad (1.51)$$

where c_1 and U_1 denote local values in the steady flow (made dimensionless through division by U_0),

$$\varphi_x = \frac{\text{const.}}{c_1 - U_1} \exp\left(ik \int^x \frac{d\xi}{c_1 - U_1}\right) \qquad (1.52)$$

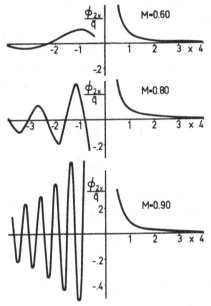

FIG. 1.4. Perturbation velocity at time $t = 2\pi n$ along the x-axis due to an acoustical source of strength $\bar{q}$ located at the origin.

The constant may be evaluated by prescribing φ_{2x} at some point, say $x = x_0$. Then, abbreviating the wave-speed $U_1 - c_1$ of the receding wave by $U_r(x)$,

$$\phi_{2x}(x, t) = \frac{U_r(x_0)}{U_r(x)} \phi_{2x}\left(x_0, t - \int_{x_0}^x \frac{d\xi}{U_r(\xi)}\right) \qquad (1.53)$$

From this solution we may draw a number of interesting conclusions.

Consider first a flow decelerating through the speed of sound so that the sonic point is to the right of $x = x_0$. Then any finite disturbance from the supersonic region must grow infinitely large as it approaches the sonic point. The linearized Eq. (1.33) cannot be valid arbitrarily close to the sonic point where $\phi_{2x} = 0(U_r)$, so that ϕ_{2x} must, of course, stay finite. However, the trend indicated by Eq. (1.53) must nevertheless be true. Hence we have reproduced the well-known results of Refs. 25, 27 and 44 that a decelerating transonic flow is unstable to small disturbances. In an accelerating flow the receding waves always move away from the sonic point and their amplitudes consequently damp out as U_r^{-1}. This is in contrast to the acoustical theory which, according to Eqs. (1.44) and (1.45) predicts an amplitude independent of x. We may infer that this property also is retained for two- and three-dimensional flows so that the receding waves damp out much faster than predicted by acoustical theory.

Finally, consider real non-linear effects which, as follows from Eq. (1.53), always must be important sufficiently close to the sonic region. A typical effect of non-linearity is that different parts of the wave get different wave speeds leading, for example, to the well known Riemann steepening of the wave-fronts. In case of a receding wave which causes a perturbation no longer small compared to that of the non-lifting flow we will have, for near-sonic flow

$$\phi_{2x} = 0(U_r) \tag{1.54}$$

Hence for the wavelength, λ_r, of the receding wave

$$\lambda_r \sim \frac{U_r}{k} = 0\left(\frac{\phi_{2x}}{k}\right) \tag{1.55}$$

The difference in speed between compression and expansion waves will be of order ϕ_{2x}. Hence the time t_r needed for a compression wave to overtake an expansion wave will be

$$t_r \sim \frac{\lambda_r}{\phi_{2x}} = 0\left(\frac{1}{k}\right) \tag{1.56}$$

The distance x_r traveled by the wave during this time is

$$x_r \sim t_r U_r = 0\left(\frac{U_r}{k}\right) \tag{1.57}$$

If we require x_r to be short compared to the reference length we arrive at the condition (1.40) above for the linearized equation (1.36) to be valid. Thus the very interesting conclusion is reached that not only will non-linearities prevent the amplitude of receding waves from becoming large, but also will cause them to damp out within a few wavelengths (in a process roughly analogous to the breaking of water waves in the surf).

Thus we have shown that the acoustical theory leads to physically inadmissible results in the transonic region. In fact, the true solution must, as the wavelength of the receding wave diminishes, approach that given by the transonic equation (1.36) according to which this wave stays in the plane $x =$ constant of its origin (see Eq. (1.46)), so that, for example, there will be no upstream influence in a slightly subsonic flow.

1.7. Generalized aerodynamic forces

The first-order equation for the pressure coefficient may be obtained from the Bernoulli equation (1.4). In isentropic flow $p/p_0 = (c^2/c_0^2)^{\gamma/(\gamma-1)}$ and hence

$$C_p = \frac{2}{\gamma M^2}\left\{\left[1 - \frac{\gamma-1}{2}M^2(2\phi_x + 2\phi_t + \phi_x^2 + \phi_y^2 + \phi_z^2)\right]^{\gamma/(\gamma-1)} - 1\right\} \quad (1.58)$$

Expanding and retaining only first-order terms gives

$$C_p = -2\phi_x - 2\phi_t - \phi_y^2 - \phi_z^2 \quad (1.59)$$

Since in the following we will only be concerned with the unsteady lifting part of the flow we will drop from now on the index 2 referring to this part of the flow. Also index 0 on U and ρ denoting free-stream values will be dropped. For an oscillating wing we set $\phi = \mathrm{Re}\{\varphi e^{ikt}\}$ and the lifting pressure difference acting on a planar wing is then given by

$$\Delta p = 2\rho U^2 e^{ikt}[\varphi_x + ik\varphi]_{z=+0} \quad (1.60)$$

One may express the final results for a set of flutter modes, flexible or rigid, in form of generalized aerodynamic force coefficients L_{ij}. Let the displacement distribution function for harmonic oscillations in a mode i, with the non-dimensional amplitude δ_i, be given by

$$h = \mathrm{Re}\{f_i(x, y)e^{ikt}\} \quad (1.61)$$

The generalized aerodynamic force coefficients are then defined as follows:

$$L_{ij} = (\tfrac{1}{2}\rho U^2 S e^{ikt}\,\delta_i)^{-1}\iint_S \Delta p_i f_j \, dx \, dy \quad (1.62)$$

where $\Delta p_i(x, y)$ is the lifting pressure difference on the wing due to the mode i.

The coefficients L_{ij} are generally complex quantities so that

$$L_{ij} = L_{ij}' + iL_{ij}'' = |L_{ij}| \exp(i\theta_{ij}) \quad (1.63)$$

For flutter calculations it is sometimes convenient to introduce sectional coefficients $l_{ij}(y) = l_{ij}' + il_{ij}'' = |l_{ij}| \exp(i\phi_{ij})$ by setting

$$L_{ij} = S^{-1}\int_{-\sigma}^{\sigma} b l_{ij} \, dy \quad (1.64)$$

where $b(y)$ now is the local chord at station y. The two-dimensional (strip theory) value of l_{ij} is denoted by $l_{ij}^{(s)}$. It is instructive, as well as convenient when introducing the three-dimensional results into flutter calculations originally planned for use of strip-theory coefficients, to define three-dimensional-flow correction coefficients c_{ij} as follows

$$c_{ij} = \Delta l_{ij}/l_{ij}^{(s)}$$

where

$$\Delta l_{ij} = l_{ij} - l_{ij}^{(s)} \quad (1.65)$$

In the following we will consider some modes connected with rigid-body wing and control surface motion. The vertical-translation and pitching modes of the wing are defined by

$$f_1 = 1 \tag{1.66}$$
$$f_2 = x \tag{1.67}$$

respectively.

Thus, L_{ij} represents total lift due to translation, L_{12} moment (about $x = 0$) due to translation, etc.

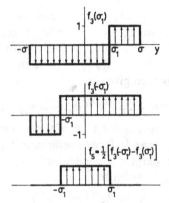

FIG. 1.5. Mode function for control surface.

There is no upstream influence in linearized transonic flow and hence, these modes could also be used for a full-span rectangular control surface by setting $x = 0$ at its leading edge and its chord equal to unity. A part-span control surface could be covered by use of the additional modes

$$f_3 = \text{sgn}\,(y - \sigma_1) \tag{1.68}$$
$$f_4 = x\,\text{sgn}\,(y - \sigma_1) \tag{1.69}$$

By a proper combination of the modes $f_1 - f_4$ any rectangular control surface configuration could be treated. The mode functions for a particular configuration will be denoted by f_5 and f_6. (The mode f_5 is incorporated so that cases with various degrees of aerodynamic balance also could be treated.) Thus in the case of a control surface of unit chord spanning from $y = -\sigma_1$ to $y = \sigma_1$ we have (see Fig. 1.5):

$$f_5 = \tfrac{1}{2}[f_3(-\sigma_1) - f_3(\sigma_1)] \tag{1.70}$$
$$f_6 = \tfrac{1}{2}[f_4(-\sigma_1) - f_4(\sigma_1)] \tag{1.71}$$

Total hinge moment is given by the coefficient L_{66} (assuming no aerodynamic balance). It is convenient to use the control surface area as reference area in this case. Another coefficient of interest is for example l_{61} which represents the spanwise distribution of lift on control surface and wing due to control surface motion.

The coefficients for rigid body motion could of course also be expressed in terms of stability derivatives. Consider first the auxiliary mode

$$\bar{f}_2 = x - a = f_2 - af_1 \tag{1.72}$$

where $x = ab$ is the location of the pitching axis. Then, denoting the generalized forces associated with f_1 and $\bar{f}_2$ by a bar, these forces can be expressed in terms of L_{ij} as follows

$$\bar{L}_{12} = L_{12} - aL_{11}$$
$$\bar{L}_{21} = L_{21} - aL_{11} \tag{1.73}$$
$$\bar{L}_{22} = L_{22} - a(L_{21} + L_{12}) + a^2 L_{11}$$

By expressing the instantaneous angle of attack, α, and the pitching angular velocity q, in terms of f_1 and $\bar{f}_2$ one finds that (Ref. 30)

$$C_{L\alpha} = -(1/k)L_{11}''$$
$$C_{L\dot{\alpha}} = (1/k^2)L_{11}'$$
$$C_{Lq} = -(1/k)L_{21}'' - (1/k^2)L_{11} \tag{1.74}$$
$$C_{L\dot{q}} = (1/k^2)L_{12}' - (1/k^3)L_{11}''$$

Similarly for moment coefficients (positive if nose-down)

$$C_{M\alpha} = (1/k)\bar{L}_{12}''$$
$$C_{M\dot{\alpha}} = -(1/k^2)\bar{L}_{12}'$$
$$C_{Mq} = (1/k)\bar{L}_{22}'' + (1/k^2)\bar{L}_{12}' \tag{1.75}$$
$$C_{M\dot{q}} = -(1/k^2)\bar{L}_{22}' + (1/k^3)\bar{L}_{12}''$$

In the following we will use the root chord of the wing as reference length for the reduced frequency and for the moment.† The $\dot{q}$-coefficients are of little practical value for the low reduced frequencies of interest for dynamic stability problems and will therefore not be further considered. The coefficients C_{Mq} and $C_{M\dot{\alpha}}$ will only be given in the combination

$$C_{Mq} + C_{M\dot{\alpha}} = (1/k)\bar{L}_{22}'' \tag{1.76}$$

which represents the total damping of the pitching motion about the fixed axis $x = ab$. Similarly, C_{Lq} and $C_{L\dot{\alpha}}$ will only be given in the combination $C_{Lq} + C_{L\dot{\alpha}}$.

---

† Current practice in the U.S.A. is to use half the aerodynamic mean chord whereas in the U.K. the full aerodynamic mean chord is commonly used. The present use is found to give shorter formulas, and it is hoped that it should not cause extra confusion.

The British notation of stability derivatives is z_w, $z_{\dot{w}}$, m_w, $m_{\dot{w}}$, z_α, $z_{\dot{\alpha}}$, m_α, $m_{\dot{\alpha}}$, where

$$z_w + ikz_{\dot{w}} = \tfrac{1}{2}L_{11}$$
$$m_w + ikm_{\dot{w}} = \tfrac{1}{2}L_{12}$$
$$z_\alpha + ikz_{\dot{\alpha}} = \tfrac{1}{2}L_{21}$$
$$m_\alpha + ikm_{\dot{\alpha}} = \tfrac{1}{2}L_{22}$$

(1.77)

By considering the flow of momentum through a control surface surrounding the body (Ref. 49) one can show that the total lift and pitching moment acting on a non-planar body are given by

$$L = \rho U^2 b^2 \left\{ \left[\int \Delta\phi \, dy \right]_{x=1} + \int_S \int \Delta\phi_t \, dx \, dy \right\}$$

(1.78)

$$M_p = \rho U^2 b^3 \left\{ (1-a) \left[\int \Delta\phi \, dy \right]_{x=1} - \int_S \int \Delta\phi \, dx \, dy + \int_S \int (x-a)\,\Delta\phi_t \, dx \, dy \right\}$$

(1.79)

where $\Delta\phi$ is the difference of ϕ between the upper and lower surfaces of the configuration. These formulas are identical with those for a planar wing given by Eqs. (1.60) and (1.62) if the ϕ_x-term is eliminated through an integration by parts.

1.8. Similarity law for unsteady transonic flow

The Mach number can be eliminated from Eq. (1.36) by means of a Prandtl–Glauert type transformation (with $\sqrt{|1 - M^2|}$, replaced by M). This leads to the following similarity law for the velocity potential of the unsteady lifting flow

$$\phi(x, y, z, t; \sigma; k; M) = M^{-1}\phi(x, y, z, t; M\sigma; k; 1)$$

(1.80)

Hence results for other transonic Mach numbers can be found by considering the $M = 1$ solution for a wing that is stretched in the spanwise direction by a factor of M. Thus, for the generalized force coefficients defined in the previous section,

$$L_{ij}(k; M; \sigma) = M^{-1}L_{ij}(k; 1; M\sigma)$$

(1.81)

In the following chapters only the case $M = 1$ will therefore be considered.

The similarity law can be used to obtain a check on the validity of the linearized equation. For, if the arguments leading to Eq. (1.36) were correct (this is of course not obvious from a mathematical point of view), the linearized subsonic and supersonic solutions should follow the similarity law when $k \gg (1 - M)$. In the two-dimensional case forces and moments, for a given k, should then be proportional to M^{-1}. To check this we have plotted in Fig. 1.6 the coefficient l_{11} (sectional lift due to translation) multiplied by M, for $M = 0.80$ (from Ref. 2), for $M = 1.0$ (from Ref. 51) and for $M = 1.11$ (from Ref. 16). As seen they all follow fairly well the same curve for large k, and the similarity law holds to within 5° in phase angle and 10 per cent in vector magnitude for

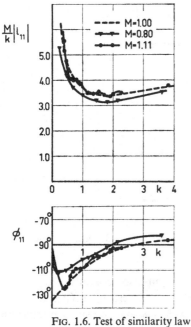

FIG. 1.6. Test of similarity law for unsteady transonic two-dimensional flow.

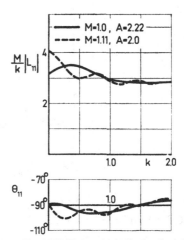

FIG. 1.7. Test of similarity law for unsteady transonic three-dimensional flow. Rectangular wing.

$k/(1 - M)$ larger than about 4–6. As pointed out in Section 1.6, the waviness of the supersonic values is due to the strong receding wave that starts at the leading edge. One would expect that this property is also exhibited by three-dimensional wings with unswept leading edges. Therefore we have plotted in Fig. 1.7 $M \cdot L_{11}$ (total lift due to translation) for a rectangular wing of $A = 2$ at $M = 1.11$ using Miles' theory† (Ref. 45) and also for a wing of $A = 2.22$ at $M = 1$ using the theory of Chapter 7. These two wings are similar according to Eq. (1.80). It is evident that a strong receding wave does indeed exist for a rectangular wing and that the frequency requirement is about the same as for the two-dimensional case. The small oscillations vs. M that is exhibited by the supersonic strip theory forces will therefore also occur for a rectangular wing of

† Actually, Miles' theory is only valid for $A \geqslant 2.08$ for this Mach number since otherwise the side edges interact. The error for $A = 2$ is certainly negligible, however.

finite aspect ratio. Since these oscillations sometimes are bothersome in flutter calculations, it is useful to know that, as was discussed in Section 1.6, they are unreal and stem from the erroneous linearization of the problem.

For a delta wing we have similarly plotted in Fig. 1.8 $M \cdot L_{11}$ for $A = 2.0$ at

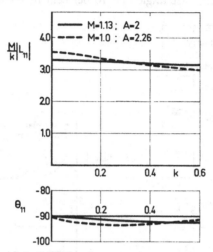

FIG. 1.8. Test of similarity law for unsteady transonic three-dimensional flow. Delta wing.

$M = 1.13$, and for $A = 2.26$ at $M = 1$.† The supersonic solution was taken from Ref. 69 and the transonic solution is the one described in Chapter 3. In this case the two solutions agree well even for low k and the small oscillations of the supersonic result do not occur. (Note the different frequency scale from Fig. 1.7, however.) Calculations for the damping in pitch given in Ref. 30 show that for low-aspect-ratio delta wings the transonic and supersonic solutions agree well for $k/(1 - M) > 1$. From Ref. 72 it is indeed found that the receding wave contribution to the pressure vanishes as $M \rightarrow 1$ in contrast to what is the case for the rectangular wing (Ref. 46).

Since the non-linear effects are associated with the receding wave one may conclude that linearized theory will be a better approximation for a wing with swept leading edges, where the receding wave contribution from the leading edge is weak, than for a wing with a straight leading edge.

1.9. Linearized boundary value problem for the planar wing at $M = 1$

With $M = 1$ Eq. (1.36) becomes

$$\phi_{yy} + \phi_{zz} - 2\phi_{xt} - \phi_{tt} = 0 \qquad (1.82)$$

† The Mach number here as for the preceding example was chosen to fit available tables.

The tangency condition, Eq. (1.28), to be applied at the projection of the wing on $z = 0$, reads

$$\phi_z = \delta(h_x + h_t) \tag{1.83}$$

Outside the wing no lift can be supported by the plane $z = 0$. Hence outside the wake

$$\phi = 0 \tag{1.84}$$

In the wake application of Eq. (1.59) gives

$$\phi_x + \phi_t = 0 \tag{1.85}$$

For a swept trailing edge the pressure must be continuous on the edge (Kutta condition). Equation (1.82) is hyperbolic and the boundary value problem is completed by postulating that ϕ vanishes for t less than a certain value and that it remains finite at large distances from the wing.

For harmonic oscillations we set

$$\phi = \mathrm{Re}\,\{\varphi e^{ikt}\} \tag{1.86}$$

$$h(x, y, t) = \mathrm{Re}\,\{f(x, y)e^{ikt}\} \tag{1.87}$$

The boundary value problem posed by Eqs. (1.82)–(1.85) then becomes

$$\varphi_{yy} + \varphi_{zz} - 2ik\varphi_x + k^2\varphi = 0 \tag{1.88}$$

$$\varphi_z(x, y, 0) = w(x, y) = \delta(f_x + ikf) \quad \text{on the wing} \tag{1.89}$$

$$\varphi(x, y, 0) = 0 \quad \text{outside the wing and wake} \tag{1.90}$$

$$\varphi_x + ik\varphi = 0 \quad \text{in the wake} \tag{1.91}$$

1.10. Transformations of the equations of motion

Equations (1.88)–(1.91) may be considered Fourier transformations with respect to time (Fourier variable $= -k$) of Eqs. (1.82)–(1.85). This line of thought will not be further pursued, however. Instead we will consider the Fourier transformation with respect to x of Eqs. (1.88)–(1.91). Fourier transformed functions are denoted by capital letters so that, for example,

$$\Phi = \mathscr{F}\{\varphi\} = \frac{1}{\sqrt{(2\pi)}} \int_{-\infty}^{\infty} \varphi e^{-iux}\, dx \tag{1.92}$$

Transformation of Eqs. (1.88)–(1.90) leads to

$$\Phi_{yy} + \Phi_{zz} + K^2\Phi = 0 \tag{1.93}$$

$$\Phi_z(y, 0) = W(y) \quad \text{for } |y| < \sigma \tag{1.94}$$

$$\Phi(y, 0) = 0 \qquad \text{for } |y| \geqslant \sigma \tag{1.95}$$

where

$$K = (2ku + k^2)^{1/2} \tag{1.96}$$

For convergence reasons we choose u and k to have arbitrarily small negative imaginary parts. Also that branch of K is chosen for which Im $(K) < 0$. To make K single-valued a cut is therefore introduced along a straight line from $u = -k/2$ to infinity with the slope $-\arg k$ as shown in Fig. 1.9. When applying the inversion formula

$$\varphi = \mathscr{F}^{-1}\{\Phi\} = \frac{1}{\sqrt{(2\pi)}} \int_{-\infty}^{\infty} e^{iux}\Phi \, du \qquad (1.97)$$

the integral should therefore be taken just below the Re (u)-axis.

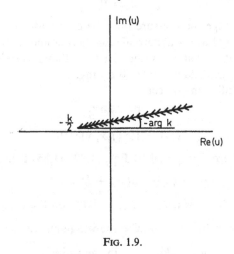

FIG. 1.9.

Since Eq. (1.88) is parabolic Laplace transformation could also have been used, but the Fourier transformation turns out to have certain advantages for some of the problems considered and is therefore used throughout.

There exists a co-ordinate transformation that leaves Eq. (1.82) invariant. We first introduce

$$x' = -2x \qquad (1.98)$$
$$t' = t - x/2 \qquad (1.99)$$

Then Eq. (1.82) becomes

$$\phi_{yy} + \phi_{zz} + 4\phi_{x't'} = 0 \qquad (1.100)$$

For this we make use of the following theorem (see Ref. 3, p. 161).

If

$$\phi = F(t', x', y, z)$$

is a solution of Eq. (1.82) then

$$\phi = \frac{1}{x'} F\left(t' + \frac{y^2 + z^2}{x'} - \frac{a^2}{x'}, \frac{ay}{x'}, \frac{az}{x'}\right)$$

is another solution.

We set $a = -2$ and return to the original variables. Then we find that if

$$\phi = P(x, y, z, t)$$

is a solution of Eq. (1.82) and we set

$$X = -1/x$$
$$Y = y/x \tag{1.101}$$
$$Z = z/x$$
$$T = t - x/2 - 1/2x - (y^2 + z^2)/2x \tag{1.102}$$

then

$$\phi = x^{-1}P(X, Y, Z, T) \tag{1.103}$$

is also a solution.

For the oscillating wing case we put $\phi = \text{Re}\,\{\varphi e^{ikt}\}$ and $P = \text{Re}\,\{\omega e^{ikT}\}$. Then φ satisfies Eq. (1.88) and

$$\varphi = x^{-1}\exp\left\{-i\frac{k}{2}[x + 1/x + (y^2 + z^2)/x]\right\}\omega(X, Y, Z) \tag{1.104}$$

is a solution, provided ω also satisfies Eq. (1.88), namely

$$\omega_{YY} + \omega_{ZZ} - 2ik\omega_X + k^2\omega = 0 \tag{1.105}$$

This transformation is used in Chapter 8 to obtain the solution for the delta wing.

CHAPTER 2

TWO-DIMENSIONAL SOLUTION

2.1. Introduction

ALTHOUGH, as we will see later on, the two-dimensional solution is of little value as an approximation to the true three-dimensional flow around practical configurations, it is nevertheless of certain theoretical interest. The two-dimensional solution can be obtained in three different ways, either directly from the transonic equation (1.88) or by taking the limit as $M \to 1$ of the sub- or super-sonic solution.

2.2. Direct approach

Consider the Fourier transformed transonic equation, Eq. (1.93), which for flow in two dimensions becomes

$$\Phi_{zz} + K^2 \Phi = 0 \qquad (2.1)$$

For $z = 0$ we require that

$$\Phi_z = W(u) \qquad (2.2)$$

Fundamental solutions of Eq. (2.1) are e^{iKz} and e^{-iKz}. Of these only the latter is finite at infinity since we have chosen $\mathrm{Im}\,(K) < 0$ (see Section 1.9). Hence the solution of Eqs. (2.1) and (2.2) is

$$\Phi = \frac{i}{K} e^{-iKz} W(u)\,\mathrm{sgn}\,z \qquad (2.3)$$

After inversion the solution is completed. Using the inversion formula, Eq. (1.97), we need the following integral

$$\mathscr{F}^{-1}\left\{\frac{i}{K} e^{-iKz}\right\} = \frac{1}{\sqrt{(2\pi)}} \int_{-\infty}^{\infty} \frac{i}{K} e^{-iKz + iux}\,du \qquad (2.4)$$

The integration path in the complex u-plane is to be taken just below the real axis. We introduce as new integration variable K. Then, since $\mathrm{Im}\,(K) < 0$, $K = -i\,|K|$ for $u < k/2$ and the integral becomes

$$\frac{i}{k\sqrt{(2\pi)}} \exp\left(-i\frac{k}{2}x\right) \int_L \exp\left(-iKz + i\frac{K^2 x}{2k}\right) dK \qquad (2.5)$$

where L is shown in Fig. 2.1.

By the further substitution $q = (K - kz/x) \sqrt{(x/2k)}$ the integral will become of the form $\int \exp(-iq^2) \, dq$, and easily evaluated to give

$$\mathscr{F}^{-1}\left\{\frac{i}{K} e^{-iKz}\right\} = \frac{i-1}{\sqrt{(2kx)}} \exp\left[-i\frac{k}{2}(x + z^2/x)\right] \qquad (2.6)$$

Thus, using the convolution theorem for Fourier integrals, we obtain for $z > 0$:

$$\varphi = \frac{i-1}{2\sqrt{(\pi k)}} \int_0^x \frac{\exp\left[-\tfrac{1}{2}ik\{x - \xi + z^2/(x - \xi)\}\right]}{\sqrt{(x - \xi)}} w(\xi) \, d\xi \qquad (2.7)$$

This solution could of course also have been obtained by distributing unsteady sources of strength we^{ikt} along the x-axis. Such a method was used by Rott (Ref. 57). Nelson and Berman (Ref. 51), who have given extensive tables of air forces on a wing-aileron combination, rederived Rott's result by using Laplace transform and, also, by taking the limit as $M \to 1$ of the supersonic solution.

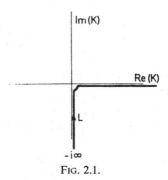

FIG. 2.1.

We will be mainly interested in the potential on the wing and the pressure distribution on it. The former is obtained simply by setting $z = 0$ in Eq. (2.7) since the integral converges uniformly in z. Application of Eq. (1.59) (and neglecting the quadratic terms) gives for the pressure on the upper surface of the wing

$$C_p = \frac{1-i}{\sqrt{(\pi k)}} e^{ikt} \left\{ x^{-1/2} w(0) \exp\left(-i\frac{k}{2}x\right) \right.$$
$$\left. + \int_0^x \frac{\exp\left[-\tfrac{1}{2}ik(x - \xi)\right]}{\sqrt{(x - \xi)}} (w_\xi + ikw) \, d\xi \right\} \qquad (2.8)$$

A square root singularity for the pressure at the leading edge occurs also for three-dimensional wings both for unswept and swept leading edges. In this respect the transonic solution is of subsonic character but not, of course, near the trailing edge since the transonic solution does not admit any upstream influence and hence is not influenced by the wake. This will be further discussed below.

2.3. Sonic limit of the supersonic solution

We will next investigate the limit as $M \to 1$ of the supersonic solution. This solution (Ref. 16) may be written, for $z = +0$,

$$\varphi = -(M^2 - 1)^{-1/2} \int_0^x e^{-i\bar{\kappa}(x-\xi)} J_0\left[\frac{\bar{\kappa}}{M}(x - \xi)\right] w(\xi)\, d\xi \qquad (2.9)$$

where

$$\bar{\kappa} = \frac{kM^2}{M^2 - 1}$$

As $M \to 1$ the argument of the Bessel function becomes infinitely large, except for $x = \xi$, and we may therefore use the asymptotic formula for J_0 over most of the wing, namely

$$J_0(z) \sim \sqrt{\left(\frac{2}{\pi z}\right)} \cos\left(z - \frac{\pi}{4}\right) = \frac{\exp(-\pi i/4)}{\sqrt{(2\pi z)}}\left(e^{iz} + ie^{-iz}\right) \qquad (2.10)$$

Hence

$$(M^2 - 1)^{-1/2} e^{-i\bar{\kappa}x} J_0\left(\frac{\bar{\kappa}}{M}x\right) \sim \frac{\exp(-\pi i/4)}{\sqrt{(2\pi kMx)}}$$

$$\times \left[\exp\left(-i\frac{kM}{M+1}x\right) + i\exp\left(-i\frac{kM}{M-1}x\right)\right] \qquad (2.11)$$

The point $x = \xi$ turns out not to give any extra contribution to the integral in the limit so that we can directly introduce Eq. (2.11) into Eq. (2.9). Thus

$$\lim_{M \to 1}\{\varphi\} = -\lim_{M \to 1}\left\{\frac{\exp(-\pi i/4)}{\sqrt{(2\pi k)}}\right.$$

$$\times \left.\int_0^x \frac{\exp[-i\{kM/(M+1)\}(x-\xi)] + i\exp[-i\{kM/(M-1)\}(x-\xi)]}{\sqrt{(x-\xi)}} w(\xi)\, d\xi\right\} \qquad (2.12)$$

The last exponential term will in the limit give no contribution to the integral (Liouville's theorem) so that the final result becomes

$$\varphi = -\frac{e^{-\pi i/4}}{\sqrt{(2\pi k)}}\int_0^x \frac{\exp[-\tfrac{1}{2}ik(x-\xi)]}{\sqrt{(x-\xi)}} w(\xi)\, d\xi \qquad (2.13)$$

i.e. identical with Eq. (2.7) for $z = 0$. As noticed by Jordan (Ref. 24), the pressure obtained from the supersonic theory, however, will not agree with the transonic value. This follows from (1.59) and (2.12) which for M very close to unity give

$$C_p \sim e^{ikt} \frac{1-i}{\sqrt{(2\pi k)}} \left\{ x^{-1/2} \left[\exp\left(-i\frac{k}{2}x\right) + i \exp\left(-i\frac{kMx}{M-1}\right) \right] w(0) \right.$$
$$\left. + \int_0^x \frac{\exp\left[-\tfrac{1}{2}ik(x-\xi)\right]}{\sqrt{(x-\xi)}} (w_\xi + ikw)\, d\xi \right\} \quad (2.14)$$

Thus the supersonic pressure distribution does not possess a limit for $M = 1$. Equation (2.14) was used for calculating the pressure distribution shown in Fig. 1.2c.

2.4. The sonic limit of the subsonic solution

The general solution for an oscillating airfoil in subsonic flow was given by Timman *et al.* (Ref. 65) as an infinite series of Mathieu functions. Although of course possible in principle, it does not seem practical, however, to use this solution for finding the limit as $M \to 1$.

Approximate solutions for high subsonic Mach numbers have been given by Burger (Ref. 6) and by Eckhaus (Ref. 10). We will here present a related solution based on a method by Schwartzschild (Ref. 60) which will be frequently used in the following and which is described in detail in Chapter 7. According to this method edge effects are calculated from the fundamental solution, P, of the wave equation,

$$P_{xx} + P_{zz} + \kappa^2 P = 0 \quad (2.15)$$

having a given value $P = F(x)$ on the positive x-axis but $P_z = 0$ on the negative x-axis. This solution may be written

$$P = \frac{1}{\pi} \int_0^\infty G(x, \xi, z) F(\xi)\, d\xi \quad (2.16)$$

It is shown in Section 7 that for $z = 0$ and $x < 0$

$$G = \sqrt{\left(\frac{-x}{\xi}\right)} \frac{e^{-i\kappa(\xi-x)}}{\xi - x} \quad (2.17)$$

For convenience we will in the following denote the distance inboard from the edge by r. As integration variable for the distance outboard from the edge we will use ρ. Hence, in the solution above, where the edge is located at the origin, we set $-x = r$ and $\xi = \rho$. Variables pertaining to a particular edge will be distinguished by a suitable index like L for leading edge and T for trailing edge.

Considering now the subsonic airfoil problem we seek a solution of the acoustical equation

$$(1 - M^2)\varphi_{xx} + \varphi_{zz} - 2ikM^2\varphi_x + k^2M^2\varphi = 0 \quad (2.18)$$

subject to the boundary conditions

$$\varphi_z(x, 0) = w(x) \quad \text{for } 0 < x < 1 \tag{2.19}$$

$$\varphi(x, 0) = 0 \qquad \text{for } x \leqslant 0 \tag{2.20}$$

$$\varphi_x(x, +0) + ik\varphi(x, +0) = 0 \qquad \text{for } x \geqslant 1 \tag{2.21}$$

If we discard the edge effects, which arise due to the boundary conditions (2.20) and (2.21), a solution $\varphi^{(0)}$ with a prescribed normal derivative w on $z = 0$ is easily found by standard methods to be

$$\varphi^{(0)} = \frac{i}{2\beta} \int_{-\infty}^{\infty} e^{i\kappa M(x-\xi)} H_0^{(2)}(\kappa R) w(\xi) \, d\xi \tag{2.22}$$

where

$$R = \sqrt{[(x - \xi)^2 + \beta^2 z^2]}$$

$$\beta = \sqrt{(1 - M^2)}$$

and

$$\kappa = kM/\beta^2$$

The distribution of w outside the airfoil is not known but can be chosen arbitrarily in this first approximation. One possibility is to set $w = 0$ outside the airfoil which leads to Kirchhoff's approximation. It has been proposed that this could be used as an approximate solution for high frequencies, or Mach numbers close to unity, but it turns out that it does not give the correct edge effects even to the first approximation.

$\varphi^{(0)}$ is generally not zero ahead of the leading edge as required by Eq. (2.20) so we need to add a solution $\psi^{(1)}$ which is equal to $-\varphi^{(0)}$ on the negative x-axis but does not change the normal derivative on the airfoil. Such a solution is obtained by aid of Schwartzschild's kernel given above. Eq. (2.18) is transformed to the wave equation by setting $\varphi = \varphi_1 \exp(i\kappa Mx)$ and $\kappa = kM/\beta^2$. Application of Eqs. (2.16) and (2.17) then gives on the airfoil

$$\psi^{(1)} = -\frac{1}{\pi} \int_0^{\infty} \sqrt{\left(\frac{r_L}{\rho_L}\right)} \frac{\exp\left[-i\{kM_1/(1 + M_1)\}(r_L + \rho_L)\right]}{r_L + \rho_L} \varphi^{(0)} \, d\rho_L \tag{2.23}$$

We now have a solution $\varphi^{(1)}$,

$$\varphi^{(1)} = \varphi^{(0)} + \psi^{(1)} \tag{2.24}$$

which satisfies the boundary conditions (2.19) and (2.20). This would be the final solution if the airfoil were infinite downstream and hence there were no influence from the wake. A special application of this solution will be considered in Section 2.5.

The pressure coefficient associated by $\varphi^{(1)}$,

$$C_p^{(1)} = -2(\varphi_x^{(1)} + ik\varphi^{(1)})e^{ikt} \tag{2.25}$$

is not zero in the wake, so we must add another solution $\bar{\psi}^{(1)}$ that gives

$$\bar{C}_p^{(1)} = -C_p^{(0)} \tag{2.26}$$

there, without altering the normal derivative on the airfoil. To find this, it is most convenient to work directly with the pressure coefficient. It is easily shown that $C_p(x, z)$, as well, is a solution of the acoustical equation (2.18). Also

$$C_{pz}(x, 0) = -2(w_x + ikw)e^{ikt}$$

so that $\bar{C}_p^{(1)}$ should have a zero normal derivative on the airfoil. Hence the solution can be obtained by use of the Schwartzschild kernel for the trailing edge and using the results given above we find that on the airfoil

$$\bar{C}_p^{(1)} = -\frac{1}{\pi} \int_1^\infty G e^{i\kappa M(x-\xi)} C_p^{(1)} \, d\xi$$

$$= -\frac{1}{\pi} \int_0^\infty \sqrt{\left(\frac{r_T}{\rho_T}\right)} \frac{\exp\left[-i\kappa(r_T + \rho_T)\right]}{r_T + \rho_T} C_p^{(1)} \, d\rho_T \tag{2.27}$$

The contribution $\bar{\psi}^{(1)}$ to the velocity potential can then be found from the formula

$$\bar{\psi}^{(1)} = -\frac{1}{2} e^{-ikt} \int_{-\infty}^x \bar{C}_p^{(1)}(u, z, t + u - x) \, du \tag{2.28}$$

Now, $\bar{\psi}^{(1)}$ is not zero ahead of the leading edge and should thus in turn be canceled. Continuing this way we obtain an infinite series for C_p which can be shown to converge for all non-zero values of k. We will describe this method much more in detail in Chapter 7 where it will be used to obtain a solution for the rectangular wing. However, for the present purpose we need only consider the first two terms, e.g.

$$C_p = C_p^{(1)} + \bar{C}_p^{(1)} \tag{2.29}$$

In calculating the limit as $M \to 1$ we first consider $C_p^{(1)}$. It is convenient to choose for $\varphi^{(0)}$ the Kirchhoff approximation. Then, by substituting in Eq. (2.22) the asymptotic expression for $H_0^{(2)}$ for large values of κR we find that

$$\varphi^{(0)} \sim \exp\left(-i\frac{kMx}{1-M}\right)(1-M)$$

for $x < 0$, whereas for $x > 0$

$$\varphi^{(0)} = [\varphi^{(0)}]_{M=1} + 0(1-M)$$

so that, through combination with Eq. (2.23) it follows that $C_p^{(1)}$ will in the limit tend to the sonic value given by Eq. (2.8). In calculating $\bar{C}_p^{(1)}$ from Eq. (2.27) we may therefore substitute in the integral the sonic value for $C_p^{(1)}$. In the limit of $M \to 1$ the main contribution in the integral comes from around $\rho_T = 0$. Hence, asymptotically

$$\bar{C}_p^{(1)} \sim -\frac{1}{\pi} C_p^{(1)}(\rho_T = 0) \int_0^\infty \sqrt{\left(\frac{r_T}{\rho_T}\right)} \frac{\exp\left[-i\{kM/(1-M)\}(r_T + \rho_T)\right]}{r_T + \rho_T} d\rho_T$$

$$= -(1 + i)C_p^{(1)}(\rho_T = 0)\left[C\left(\frac{kMr_T}{1-M}\right) - iS\left(\frac{kMr_T}{1-M}\right) - \frac{1}{2} + \frac{i}{2}\right] \quad (2.30)$$

where C and S denote the Fresnal cosine and sine integrals, respectively. Thus the final asymptotic formula obtained for the pressure distribution at high subsonic Mach numbers is (going back to the original co-ordinate system)

$$C_p \sim [C_p(x)]_{M=1} + [C_p(1)]_{M=1}F(1 - x) \quad (2.31)$$

where

$$F(x) = (1 + i)\left[G\left(\frac{kMx}{1-M}\right) - iS\left(\frac{kMx}{1-M}\right) - \frac{1}{2} + \frac{i}{2}\right] \quad (2.32)$$

A similar, but not identical, result was obtained empirically by Jordan (Ref. 24). Note that the pressure converges to the transonic result at $M = 1$ except at the trailing edge, in contrast to the supersonic pressure, Eq. (2.14). The second term in Eq. (2.31) represents a receding wave that starts at the trailing edge. Apparently, this is much weaker in the subsonic than in the supersonic case due to the different boundary condition in the region $x \geq 1$ from that in the region $x < 0$. However, if w is discontinuous, as it will be at the leading edge of a control surface, C_p will in the limit $M \to 1$ have a non-vanishing receding-wave part like in the supersonic case, so that the subsonic pressure converges to the sonic one only if w is continuous on the airfoil. Equation (2.31) was used in calculating the pressure distribution shown in Fig. 1.2a.

We emphasize again that the asymptotic formulas (2.14) and (2.31) are only of theoretical interest since, as shown in Chapter 1, the correct linearized equation to use near $M = 1$ is the transonic one, Eq. (1.36) and not the acoustical equation (1.7).

2.5. Strip theory for a wing with swept leading edge

We conclude this chapter by mentioning the special strip theory developed in Ref. 32 for a wing with a swept, subsonic leading edge but supersonic or sonic trailing edge. Let the leading edge sweep be Λ. As new independent variables we introduce

$$x_1 = x - y \tan \Lambda$$

$$y_1 = y$$

$$z_1 = z \cos \Lambda \quad (2.33)$$

Hence x_1 is measured in the free-stream direction from the leading edge at the spanwise station y_1. If derivatives in the y_1-direction are neglected the differential equation to solve becomes

$$(1 - M_N)\varphi_{x_1x_1} + \varphi_{z_1z_1} - 2ikM_N{}^2\varphi_{x_1} + k^2M_N{}^2\varphi = 0 \qquad (2.34)$$

where $M_N = M\cos\Lambda$. This is identical to Eq. (2.18) for two-dimensional subsonic flow with the Mach number M_N. The present problem, however, differs in the boundary conditions since, due to the supersonic trailing edge, the chord should be assumed infinite in the downstream direction. Hence Eqs. (2.23) and (2.24) are directly applicable so that the solution is obtained as

$$\varphi = \varphi^{(0)} - \frac{1}{\pi}\int_0^\infty \sqrt{\left(\frac{r_L}{\rho_L}\right)}\ \frac{\exp\left[-i\{kM_N/(1 + M_N)\}(r_L + \rho_L)\right]}{r_L + \rho_L}\varphi^{(0)}\,d\rho_L \qquad (2.35)$$

In calculating $\varphi^{(0)}$, w should be multiplied by $\cos\Lambda$ in view of Eq. (2.33). In Ref. 32 where a different method of solution was used, expressions for forces and moments for translational and pitching oscillations were given. It is interesting to notice that these coefficients do not oscillate when M approaches unity from above like the ordinary strip theory ones. For $M = 1$ the special strip theory just described is only applicable to wings with unswept trailing edges. Note, however, that it is consistent with the linearization process to neglect the sweep whenever $|1 - M_N|/|1 - M_L| = 0(1)$ (see Chapter 1).

CHAPTER 3

LOW ASPECT RATIO WINGS OF TRIANGULAR AND RELATED PLANFORMS

3.1. Introduction

As ALREADY noticed in Chapter 1 the linearized differential equation for unsteady transonic flow degerates to the Laplace equation in the cross-flow $(y, z\text{-})$ plane as $k \to 0$. This simple equation is the basis of the well-known slender-body approximation which has been successfully used in a number of steady-flow problems. Since most wings of practical interest for transonic airplanes are low aspect ratio a natural method of attack for the unsteady problem would be to utilize the slender-wing solution as a starting-point. Such a method was developed by Adams and Sears (Ref. 1) for sub- and super-sonic steady flow and its extension to unsteady transonic flow is described in this chapter and in Chapter 5.

3.2. Solution by iteration

Consider the Fourier transformed differential equation (1.93), namely

$$\Phi_{yy} + \Phi_{zz} + K^2\Phi = 0 \tag{3.1}$$

where

$$K = (2ku + k^2)^{1/2}$$

A solution representing a doublet distribution is

$$\Phi = -\frac{iK}{2} \int_{-\sigma}^{\sigma} \frac{z}{\tilde{r}} H_1^{(2)}(K\tilde{r})\Phi(\eta + 0) \, d\eta \tag{3.2}$$

where

$$\tilde{r} = \sqrt{[(y - \eta)^2 + z^2]} \tag{3.3}$$

The integration limits follows from the fact that the loading, i.e. φ on $z = 0$, is zero outside the side edges of the wing. Upon inversion Eq. (3.2) leads to the integral equation for the oscillating-surface problem at $M = 1$. An asymptotic expansion valid for low values of σ is obtained by expanding the Hankel function for small values of its argument, which leads to

$$\Phi = \frac{1}{\pi} \int_{-\sigma}^{\sigma} \frac{z\Phi \, d\eta}{\tilde{r}^2} + z\left\{\frac{K^2}{4\pi}(1 - \lambda)\int_{-\sigma}^{\sigma} \Phi \, d\eta - \frac{K^2}{2\pi} \int_{-\sigma}^{\sigma} \Phi \ln \tilde{r} \, d\eta \right.$$

$$\left. - \frac{K^4}{32\pi}\left(\frac{5}{2} - \lambda\right) \int_{-\sigma}^{\sigma} \Phi\tilde{r}^2 \, d\eta + \frac{K^4}{16\pi} \int_{-\sigma}^{\sigma} \Phi\tilde{r}^2 \ln \tilde{r} \, d\eta \right\} + 0[(\sigma K)^6\Phi] \tag{3.4}$$

32

where

$$\lambda = 2\gamma + \pi i + 2 \ln (K/2) \qquad (3.5)$$

and

$$\gamma = \text{Euler's constant} = 0.5772 \ldots$$

(In Eq. (3.4) and in the following the argument of $\Phi(\eta, +0)$ is left out for shortness).

The normal derivative is prescribed for φ. We therefore differentiate Eq. (3.4) with respect to z and let $z \to 0$. This gives

$$\Phi_z = -\frac{1}{\pi} \oint_{-\sigma}^{\sigma} \frac{\Phi_\eta \, d\eta}{y - \eta} + \frac{K^2}{4\pi} \left[(1 - \lambda) \int_{-\sigma}^{\sigma} \Phi \, d\eta - 2 \int_{-\sigma}^{\sigma} \ln |y - \eta| \Phi \, d\eta \right]$$

$$- \frac{K^4}{32\pi} \left[\left(\frac{5}{2} - \lambda \right) \int_{-\sigma}^{\sigma} (y - \eta)^2 \Phi \, d\eta - 2 \int_{-\sigma}^{\sigma} (y - \eta)^2 \ln |y - \eta| \Phi \, d\eta \right] \quad (3.6)$$

By inverting the Fourier transforms and introducing the tangency condition on the wing, Eq. (1.89), we obtain

$$w = -\frac{1}{\pi} \oint_{-s}^{s} \frac{\varphi_\eta \, d\eta}{y - \eta} + \frac{ik}{2\pi} D \left\{ \left(\ln \frac{k}{2} + \gamma + \frac{\pi i}{2} - 1 \right) \int_{-s}^{s} \varphi \, d\eta \right.$$

$$+ 2 \int_{-s}^{s} \ln |y - \eta| \varphi \, d\eta - D \int_0^x \exp \left[-i \frac{k}{2} (x - \xi) \right] \ln (x - \xi) \, d\xi$$

$$\times \left[\int_{-s}^{s} \varphi \, d\eta \right] \right\} - \frac{k^2}{8\pi} D^2 \left\{ \left(\ln \frac{k}{2} + \gamma + \frac{\pi i}{2} - \frac{5}{2} \right) \right.$$

$$\times \int_{-s}^{s} (y - \eta)^2 \varphi \, d\eta + 2 \int_{-s}^{s} (y - \eta)^2 \ln |y - \eta| \varphi \, d\eta$$

$$- D \int_0^x \exp \left[-i \frac{k}{2} (x - \xi) \right] \ln (x - \xi) \, d\xi \left[\int_{-s}^{s} (y - \eta)^2 \varphi \, d\eta \right] \right\}$$

$$(3.7)$$

where $s(x)$ is the local semispan and $D = \partial/\partial x + ik/2$. Here we have assumed that the integrals over η, together with their first and second derivatives, are continuous functions of x. We may simplify the expressions by writing Eq. (3.7) symbolically as follows:

$$w = -\frac{1}{\pi} \oint_{-s}^{s} \frac{\varphi_\eta \, d\eta}{y - \eta} + \mathscr{P}^{(1)}\{\varphi\} + \mathscr{P}^{(2)}\{\varphi\}$$

where $\mathscr{P}^{(1)}$ and $\mathscr{P}^{(2)}$ stand for the combined differential and integral operators defined in Eq. (3.7) by the first and second curly brackets, respectively. The operator $\mathscr{P}^{(1)}$ causes an increase in the order of magnitude by a factor of $k\sigma^2 \ln k\sigma^2$, and $\mathscr{P}^{(2)}$ by a factor of $k^2\sigma^4 \ln k\sigma^2$. If all higher order terms in Eq. (3.8) are neglected we obtain the unsteady-flow counterpart of Jones' (Ref. 23) slender-wing theory. This extension to unsteady flow was first given by Garrick (Ref. 17).

An approximate solution of Eq. (3.8) can be obtained by iteration following Adams and Sears (Ref. 1). We set

$$\varphi = \varphi^{(1)} + \varphi^{(2)} + \varphi^{(3)} + \dots \tag{3.9}$$

where $\varphi^{(1)}$ is the slender-wing theory value and $\varphi^{(2)}$, $\varphi^{(3)}$. . . the higher-order terms provided by the present theory. Substituting φ as given by Eq. (3.9) into Eq. (3.8) and equating terms of like order gives

$$w = -\frac{1}{\pi} \oint_{-s}^{s} \frac{\varphi_\eta^{(1)} \, d\eta}{y - \eta} \tag{3.10}$$

$$0 = -\frac{1}{\pi} \oint_{-s}^{s} \frac{\varphi_\eta^{(2)} \, d\eta}{y - \eta} + \mathscr{P}^{(1)}\{\varphi^{(1)}\} \tag{3.11}$$

$$0 = -\frac{1}{\pi} \oint_{-s}^{s} \frac{\varphi_\eta^{(3)} \, d\eta}{y - \eta} + \mathscr{P}^{(1)}\{\varphi^{(2)}\} + \mathscr{P}^{(2)}\{\varphi^{(1)}\} \tag{3.12}$$

These integral equations may all be written as

$$w^{(n)} = -\frac{1}{\pi} \oint_{-s}^{s} \frac{\varphi_\eta^{(n)} \, d\eta}{y - \eta} \tag{3.13}$$

where

$$w^{(1)} = w \tag{3.14}$$

$$w^{(2)} = -\mathscr{P}^{(1)}\{\varphi^{(1)}\} \tag{3.15}$$

$$w^{(3)} = -\mathscr{P}^{(1)}\{\varphi^{(2)}\} - \mathscr{P}^{(2)}\{\varphi^{(1)}\} \tag{3.16}$$

The solution of Eq. (3.13) is well known (Ref. 61). For the present problem $\varphi^{(n)}(s, 0) = \varphi^{(n)}(-s, 0) = 0$ and the solution is given by

$$\varphi_y^{(n)} = \frac{1}{\pi\sqrt{(s^2 - y^2)}} \oint_{-s}^{s} \frac{w^{(n)}\sqrt{(s^2 - \eta^2)}}{y - \eta} \, d\eta \tag{3.17}$$

The problem is thus reduced to the evaluation of a number of fairly simple integrals. The iteration breaks down, however, if $\int_{-s}^{s} \varphi^{(1)} \, d\eta$ or its first two

x-derivatives are discontinuous since then $\varphi^{(2)}$ or $\varphi^{(3)}$ becomes infinite (due to the double integral terms in Eq. (3.7)). Hence the solution requires that the leading edge slope and curvature are continuous and that w and its first two derivatives have no discontinuities. Furthermore, the wing has to be pointed. Methods to remove the limitations are considered in Chapters 4 and 8.

3.3. Solutions for a delta wing

The simplest planform which fulfils the above requirements is the triangular one. We give below results for translational and pitching oscillations. Without loss of generality we may set the amplitude equal to unity. For translational oscillations straightforward application of the formulas in the previous section then gives for the potential distribution on the upper surface of the wing

$$\varphi^{(1)} = -ik\sqrt{(s^2 - y^2)} \tag{3.18}$$

$$\varphi^{(2)} = \frac{k^2\sigma^2}{8}\sqrt{(s^2 - y^2)}\left[\frac{2iky^2}{3\sigma^2} + (4x + ikx^2)(v + \ln x) + \frac{5ikx^2}{12} + 0(k^2)\right] \tag{3.19}$$

$$\varphi^{(3)} = \frac{ik^3\sigma^4}{32}\sqrt{(s^2 - y^2)}\left\{\frac{y^2}{\sigma^2}\left[4(v + \ln x) + \frac{2}{3}\right] + 12x^2\left[(v + \ln x)^2 + \frac{\pi^2}{6}\right]\right.$$
$$\left. + 14x^2(v + \ln x) - \frac{38}{3}x^2 + 0(k)\right\} \tag{3.20}$$

where

$$v = \gamma - \ln\frac{8}{k\sigma^2} + \frac{\pi i}{2} \tag{3.21}$$

In obtaining these results we have expanded integrals like

$$\int_0^x \exp\left[-i\frac{k}{2}(x - \xi)\right]\ln(x - \xi)\,d\xi\left[\int_{-s}^s \varphi^{(1)}\,d\eta\right]$$

in powers of k so that the results are valid only for low k.

The results for pitching oscillations are:

$$\varphi^{(1)} = -(1 + ikx)\sqrt{(s^2 - y^2)} \tag{3.22}$$

$$\varphi^{(2)} = -\frac{ik\sigma^2}{4}\sqrt{(s^2 - y^2)}\left\{\frac{iky^2}{3\sigma^2}(3 + ikx)\right.$$
$$\left. + \left(2x + \frac{7}{2}ikx^2 - \frac{k^2}{2}x^3\right)(v + \ln x) + \frac{5}{4}ikx^2 + 0(k^2)\right\} \tag{3.23}$$

$$\varphi^{(3)} = \frac{k^2\sigma^4}{32}\sqrt{(s^2 - y^2)}\left\{\frac{y^2}{\sigma^2}\left[4(v + \ln x) + \frac{2}{3}\right] + 12x^2\left[(v + \ln x)^2 + \frac{\pi^2}{6}\right]\right.$$
$$\left. + 14x^2(v + \ln x) - \frac{38}{3}x^2 + 0(k)\right\} \tag{3.24}$$

Results for rolling and bending modes were also given in Ref. 30.

3.4. Generalized force coefficients

We list below the results for the four total-force coefficients $L_{11} - L_{22}$ defined in Section 1.7. These are most easily obtained by carrying out the y-integration first. It is convenient to denote a coefficient corresponding to a particular approximation by a superscript in the same manner. Thus $L_{ij} = L_{ij}^{(1)} + L_{ij}^{(2)} + L_{ij}^{(3)}$ and it is found that

$$\frac{2}{\pi A} L_{11}^{(1)} = -ik + \frac{1}{3}k^2 \tag{3.25}$$

$$\frac{2}{\pi A} L_{11}^{(2)} = \frac{k^2\sigma^2}{4}\left[(2 + ik)v + \frac{3ik}{8} + 0(k^2)\right] \tag{3.26}$$

$$\frac{2}{\pi A} L_{11}^{(3)} = \frac{ik^3\sigma^4}{16}\left[6\left(v^2 + \frac{\pi^2}{6}\right) + \frac{15}{4}v - \frac{25}{4} + 0(k)\right] = ik\left[\frac{2}{\pi A}L_{21}^{(3)}\right] \tag{3.27}$$

$$\frac{2}{\pi A} L_{12}^{(1)} = -\frac{2}{3}ik + \frac{1}{4}k^2 \tag{3.28}$$

$$\frac{2}{\pi A} L_{12}^{(2)} = \frac{k^2\sigma^4}{4}\left[\left(\frac{3}{2} + \frac{4ik}{5}\right)v + \frac{1}{2} + \frac{17}{50}ik + 0(k^2)\right] \tag{3.29}$$

$$\frac{2}{\pi A} L_{22}^{(3)} = \frac{ik^3\sigma^4}{16}\left[\frac{24}{5}\left(v^2 + \frac{\pi^2}{6}\right) + \frac{263}{100}v - \frac{1199}{250} + 0(k)\right] = ik\left[\frac{2}{\pi A}L_{22}^{(3)}\right] \tag{3.30}$$

$$\frac{2}{\pi A} L_{21}^{(1)} = -1 - \frac{4}{3}ik + \frac{1}{4}k^2 \tag{3.31}$$

$$\frac{2}{\pi A} L_{21}^{(2)} = -\frac{ik\sigma^2}{4}\left[(2 + 4ik)v + \frac{11}{8}ik + 0(k^2)\right] \tag{3.32}$$

$$\frac{2}{\pi A} L_{22}^{(1)} = -\frac{2}{3} - ik + \frac{1}{5}k^2 \tag{3.33}$$

$$\frac{2}{\pi A} L_{22}^{(2)} = -\frac{ik\sigma^2}{4}\left[\left(\frac{3}{2} + \frac{16}{5}ik\right)v + \frac{1}{8} + \frac{63}{50}ik + 0(k^2)\right] \tag{3.34}$$

As in preceding section only lower order terms in k are retained so that the results are valid only for low values of k. In Fig. 3.1 are shown results for L_{11} for some values of aspect ratio and reduced frequencies. At $A = 2.5$ the convergence is poor even for low k. Taking $A\sqrt{k}$ as the important parameter for the convergence, the second-order solution is apparently in error more than 10 per cent in vector magnitude and 5° in phase angle for $A\sqrt{k} > 1.6$. The third-order solution is probably unreliable for $A\sqrt{k} > 2$.

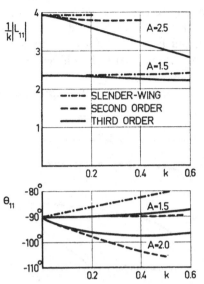

FIG. 3.1. Coefficient of total lift due to translation for low aspect ratio delta wings.

3.5. Stability derivatives for the delta wing

Casting the results of the preceding section in form of stability derivatives and retaining only terms of lower order in k gives

$$\frac{2}{\pi A} C_{L\alpha} = 1 - \frac{\pi k A^2}{64} \tag{3.35}$$

$$\frac{2}{\pi A}(C_{Lq} + C_{L\dot\alpha}) = \frac{4}{3} - a - \frac{A^2}{64}\Big[2\bar{v} + \pi k(2-a)\Big] + \frac{\pi k A^4}{1024}\Big(6\bar{v} - \frac{15}{8}\Big) \tag{3.36}$$

$$\frac{2}{\pi A} C_{M\alpha} = -\frac{2}{3} + a + \frac{\pi k A^2}{64}\Big(\frac{3}{4} - a\Big) \tag{3.37}$$

$$\frac{2}{\pi A}(C_{Mq} + C_{M\dot\alpha}) = -(1-a)^2 + \frac{A^2}{64}\Big[2\bar{v}\Big(\frac{3}{4} - a\Big) - \frac{1}{8}$$
$$+ \pi k\Big(\frac{8}{5} - \frac{11}{4}a + a^2\Big)\Big] - \frac{\pi k A^4}{4096}\Big[6\Big(\frac{4}{5} - a\Big)\bar{v} - \frac{263}{200} + \frac{15}{8}a\Big] \tag{3.38}$$

where

$$\bar{v} = \ln\frac{128}{kA^2} - \gamma \tag{3.39}$$

Note that each stability derivative contains terms of both odd and even powers of k, in contrast to sub- or super-sonic derivatives which contain only even powers. Hence, for consistency, terms of order k should be retained in the stiffness derivatives, even if all higher order terms in the damping derivatives were neglected, since damping forces are of order k. In the expressions above we have retained terms of higher order in the damping derivatives than what corresponds to those retained in the stiffness derivatives since the former are believed to be of greater importance. Also the stiffness derivatives seem to vary less rapidly with k so that one can probably for most practical cases neglect all but the lowest order (slender-body) terms in the expressions for $C_{L\alpha}$ and $C_{M\alpha}$.

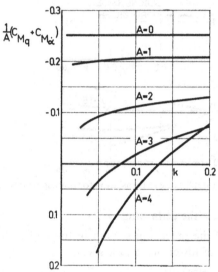

Fig. 3.2. Damping in pitch of delta wings oscillating about $a = 0.60$.

Sonic stability derivatives for a delta wing were first given by Mangler (Ref. 43) by use of a completely different method (although his original results (Ref. 42) were in error). The Adams–Sears method was used in Refs. 30 and 22 and also by Miles (Ref. 49).

Some numerical results for the derivative of total damping-in-pitch, $C_{Mq} + C_{M\alpha}$, are shown in Fig. 3.2. It is evident from the results that a delta wing of large aspect ratio may have negative pitch-damping, i.e. be susceptible to unstable short-period pitching oscillations.

3.6. Influence of planform on damping-in-pitch

If only lowest order terms are retained the quasi-steady derivatives $C_{L\alpha}$ and $C_{M\alpha}$ are given by slender-wing theory and hence for a pointed wing with unswept trailing edge, but otherwise of arbitrary planform:

$$C_{L\alpha} = \frac{\pi}{2} A \qquad (3.40)$$

$$C_{M\alpha} = -\frac{\pi A}{2}\left[(1 - a) - \sigma^{-2}\int_0^1 s^2 \, dx\right] \qquad (3.41)$$

For the unsteady derivatives we obtain after some calculations

$$\frac{2}{\pi A}(C_{Lq} + C_{L\dot{\alpha}}) = 1 - a + \sigma^{-2}\int_0^1 s^2\, dx - \frac{1}{4}\left[2\sigma s_x(1)(\bar{v}+1) + I(1)\right]\quad (3.42)$$

$$\frac{2}{\pi A}(C_{Mq} + C_{M\dot{\alpha}}) = -(1-a)^2 + \frac{1}{8}\left\{[4(1-a)\sigma s_x(1) - \sigma^2](\bar{v}+1) - \frac{1}{2}\sigma^2\right.$$
$$\left. + 2(1-a)I(1) - 2\sigma^{-2}\int_0^1 s^2 I(x)\, dx\right\}\quad (3.43)$$

where

$$\bar{v} = \frac{8}{k\sigma^2} - \gamma$$

and

$$I(x) = \frac{\partial^2}{\partial x^2}\int_0^x \ln(x - \xi)s^2(\xi)\, d\xi$$

By inspection it is seen that the pitch-damping will be high if $s_x(1)$ is low. Hence it should be favorable to the damping to round off the planform so that

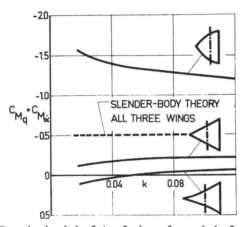

FIG. 3.3. Damping in pitch of $A = 2$ wings of curved planform with $a = 0.60$.

the leading edge slope is small near the trailing edge. This is demonstrated in Fig. 3.3 where we have considered three wings, all of aspect ratio two, defined by

(1) $s = \sigma(2x - x^2)$ (convex leading edge)   (3.44)

(2) $s = \sigma x$ (delta wing)   (3.45)

(3) $s = \frac{1}{2}\sigma(x + x^2)$ (concave leading edge)   (3.46)

According to slender-wing theory all three wings should have

$$C_{Mq} + C_{M\alpha} = -\frac{\pi A}{2}(1 - a)^2$$

(see Eq. (3.43).) With the second-order term in A included, however, they differ considerably as shown in Fig. 3.3. The convex wing has a damping that is more than six times that of the delta, whereas the concave wing has zero or slightly negative damping. Kennet *et al.* (Ref. 26) have shown that these properties extend into the low supersonic region.

CHAPTER 4

THE LOW ASPECT RATIO RECTANGULAR WING

4.1. Introduction

THE iteration method of Adams and Sears which was used in the preceding chapter only works for a class of wings which are pointed and whose leading edges are smooth functions of x. An important planform not belonging to this class is the rectangular one.

The low aspect ratio rectangular wing at incidence in steady super-sonic flow was first treated by Stewartson (Ref. 64). Miles (Ref. 46) extended the theory to unsteady flow and also gave some result for stability derivatives at $M = 1$. His technique was to Laplace transform the equations of motion in the free-stream direction and then solve the resulting two-dimensional diffraction problem around a finite strip in terms of Mathieu functions. The transformed result was then expanded for small values of the aspect ratio and the corresponding (asymptotic) inversion found. Only first-order terms in frequency were retained, appropriate to the calculation of stability derivatives.

Due to the complexity of the Mathieu functions Miles' method does not seem very practical to use for the calculation of flutter derivatives since then higher order terms in frequency, as well as solutions for elastic modes, are needed. In the method described below, which was given in Ref. 37 the velocity potential is therefore expanded in a Glauert series in the spanwise direction instead of in a Mathieu function series as in Miles' method.

4.2. Approximate solution

As in the preceding chapter the starting-point is the doublet-distribution solution of the Fourier transformed potential equation, namely

$$\Phi = -\frac{iK}{2} \int_{-\sigma}^{\sigma} \frac{z}{\tilde{r}} H_1^{(2)}(K\tilde{r}) \, \Phi(\eta, +0) \, d\eta \qquad (4.1)$$

where

$$K = (2ku + k^2)^{1/2}$$

and

$$\tilde{r} = \sqrt{[(y - \eta)^2 + z^2]}$$

Applying the transformed tangency condition, Eq. (1.94), leads to the following integral equation for the determination of $\Phi(\eta, +0)$:

$$W(y) = -\frac{iK}{2} \lim_{z \to 0} \left\{ \frac{\partial}{\partial z} \int_{-\sigma}^{\sigma} \frac{z}{\tilde{r}} H_1^{(2)}(K\tilde{r}) \, \Phi(\eta, +0) \, d\eta \right\} \qquad (4.2)$$

where $W(y)$ is the Fourier transform of the normal velocity on the wing.

To obtain the desired low aspect ratio solution we first expand the Hankel function for small values of its argument. Then the operations on the right hand side of Eq. (4.2) may be carried out with the result that

$$W(y) = -\frac{1}{\pi}\oint_0^\pi \frac{\Phi_\eta\, d\eta}{y-\eta} + \frac{K^2}{4\pi}(1-\lambda)\int_{-\sigma}^\sigma \Phi\, d\eta - \frac{K^2}{2\pi}\int_{-\sigma}^\sigma \ln|y-\eta|\Phi\, d\eta$$

$$-\frac{K^4}{32\pi}\left(\frac{5}{2}-\lambda\right)\int_{-\sigma}^\sigma (y-\eta)^2\, \Phi\, d\eta + \frac{K^4}{16\pi}\int_{-\sigma}^\sigma (y-\eta)^2 \ln|y-\eta|\, \Phi\, d\eta$$

$$+\frac{K^6}{768\pi}\left(\frac{10}{3}-\lambda\right)\int_{-\sigma}^\sigma (y-\eta)^4\, \Phi\, d\eta - \frac{K^6}{384\pi}\int_{-\sigma}^\sigma (y-n)^4 \ln|y-\eta|\, \Phi\, d\eta$$

$$+ 0[\sigma^8 K^8 \ln(K\sigma)] \tag{4.3}$$

where

$$\lambda = 2\gamma + \pi i + 2\ln(K/2)$$

and

$$\gamma = \text{Euler's constant} = 0.5772\ldots$$

(In Eq. (4.3) the argument of $\Phi(\eta, +0)$ is left out for shortness.)

Now we introduce in Eq. (4.3) the Glauert variables

$$y = \sigma\cos\psi; \quad \eta = \sigma\cos\theta \tag{4.4}$$

which gives

$$\sigma W(\psi) = \frac{1}{\pi}\oint_0^\pi \frac{\Phi_\theta\, d\theta}{\cos\psi - \cos\theta} + \frac{\varepsilon}{\pi}(1-\lambda-2\ln\sigma)\int_0^\pi \Phi\sin\theta\, d\theta$$

$$-\frac{2\varepsilon}{\pi}\int_0^\pi \ln|\cos\psi-\cos\theta|\,\Phi\sin\theta\, d\theta - \frac{\varepsilon^2}{2\pi}\left(\frac{5}{2}-\lambda-2\ln\sigma\right)$$

$$\times\int_0^\pi(\cos\psi-\cos\theta)^2\,\Phi\sin\theta\, d\theta + \frac{\varepsilon^2}{\pi}\int_0^\pi(\cos\psi-\cos\theta)^2$$

$$\times\ln|\cos\psi-\cos\theta|\,\Phi\sin\theta\, d\theta + \frac{\varepsilon^3}{12\pi}\left(\frac{10}{3}-\lambda-2\ln\sigma\right)$$

$$\times\int_0^\pi(\cos\psi-\cos\theta)^4\,\Phi\sin\theta\, d\theta$$

$$-\frac{\varepsilon^3}{6\pi}\int_0^\pi(\cos\psi-\cos\theta)^4\ln|\cos\psi-\cos\theta|\,\Phi\sin\theta\, d\theta + 0(\varepsilon^4\lambda) \tag{4.5}$$

where

$$\varepsilon = \frac{\sigma^2 K^2}{4} = \frac{\sigma^2 k}{2}\left(s+\frac{k}{2}\right)$$

To obtain an approximate solution we form from Eq. (4.5) the weighted integrals

$$\int_0^\pi \sin \psi \sin n\, \psi\, W(\psi)\, d\psi = W_n \tag{4.6}$$

For the calculation of these the following two formulas are useful

$$\frac{1}{\pi} \oint_0^\pi \frac{\sin n\, \psi \sin \psi}{\cos \psi - \cos \theta}\, d\psi = -\cos n\, \theta \tag{4.7}$$

$$\ln |\cos \psi - \cos \theta| = -\ln 2 - 2 \sum_{r=1}^\infty \frac{1}{r} \cos r\, \psi \cos r\, \theta \tag{4.8}$$

By use of these plus the orthogonality relation for the trigonometric functions the integrals may all be given in terms of the weighted potentials $\Phi_n(u)$, where

$$\sigma\, \Phi_n = \int_0^\pi \Phi(\theta) \sin n\, \theta\, d\theta \tag{4.9}$$

For $n = 1$ the integration gives the result

$$
\begin{aligned}
-W_1 = {} & \left[1 - \frac{\varepsilon}{2}\left(\frac{3}{2} - \nu\right) + \frac{\varepsilon^2}{8}\left(\frac{7}{6} - \nu\right) - \frac{5\varepsilon^3}{192}\left(\frac{91}{60} - \nu\right) \right]\Phi_1 \\
& + \frac{\varepsilon}{4}\left[1 + \frac{\varepsilon}{4}\left(\frac{1}{4} - \nu\right) - \frac{3\varepsilon^2}{32}\left(\frac{51}{45} - \nu\right) \right]\Phi_3 \\
& - \frac{\varepsilon^2}{192}\left[1 - \frac{\varepsilon\nu}{2} \right]\Phi_5 + \frac{\varepsilon^3}{11{,}520}\Phi_7 + 0(\varepsilon^4\nu)
\end{aligned} \tag{4.10}
$$

where

$$\nu = \ln(\varepsilon/4) + 2\gamma + \pi i$$

In order to obtain Φ_1 correct up to terms of order ε^4 from Eq. (4.10) we thus need to know Φ_3 to order ε^3, Φ_5 to order ε^2, and Φ_7 to order ε. The weighted integral for $n = 7$ gives,

$$\Phi_7 = -\frac{1}{7} W_7 + 0(\varepsilon) \tag{4.11}$$

which is, of course, the slender-wing-theory value.

For $n = 5$ we get

$$-\frac{1}{5} W_5 = \frac{\varepsilon}{40} \Phi_3 + \left(1 - \frac{\varepsilon}{24}\right)\Phi_5 + \frac{\varepsilon}{60} \Phi_7 + 0(\varepsilon^2) \tag{4.12}$$

Using Eq. (4.11) and solving for Φ_5 gives, neglecting terms of order ε^2

$$\Phi_5 = -\frac{\varepsilon}{40} \Phi_3 - \frac{1}{5} W_5\left(1 - \frac{\varepsilon}{24}\right)^{-1} + \frac{\varepsilon}{420} W_7 + 0(\varepsilon^2) \tag{4.13}$$

Similarly for $n = 3$

$$-\frac{1}{3} W_3 = \frac{\varepsilon}{12}\left[1 + \frac{\varepsilon}{4}\left(\frac{1}{4} - v\right)\right]\Phi_1 + \left(1 - \frac{\varepsilon}{8} - \frac{3\varepsilon^2}{160}\right)\Phi_3$$

$$+ \frac{\varepsilon}{24}\left(1 + \frac{\varepsilon}{12}\right)\Phi_5 - \frac{\varepsilon^2}{2880}\Phi_7 + 0(\varepsilon^3) \quad (4.14)$$

Equations (4.11) and (4.13) are used to eliminate Φ_7 and Φ_5 so that Φ_3 may be solved from Eq. (4.14) giving

$$\Phi_3 = -\frac{\varepsilon[(1 + \varepsilon/4)(1/4 - v)]}{12(1 - \varepsilon/8)}\Phi_1 - \frac{W_3}{3(1 - \varepsilon^2/8 - 19\varepsilon^2/960)}$$

$$+ \frac{\varepsilon(1 + \varepsilon/12)}{120(1 - \varepsilon/24)(1 - \varepsilon/8)}W_5 - \frac{\varepsilon^2}{6720}W_7 + 0(\varepsilon^3) \quad (4.15)$$

Note that terms like $(1 - \varepsilon/8)^{-1}$, although formally $(1 - \varepsilon/8)^{-1} = 1 + \varepsilon/8 + 0(\varepsilon^2)$, have been left unexpanded in Eqs. (4.13) and (4.15) in order to retain the poles of Φ_n (see below).

Now Φ_3, Φ_5, and Φ_7 can be eliminated from Eq. (4.10) and Φ_1 solved. The result may then be expressed as follows

$$\Phi_1 = H_{11}W_1 + H_{13}W_3 + H_{15}W_5 + H_{17}W_7 + 0(\varepsilon^4 v) \quad (4.16)$$

where

$$D_1 H_{11} = -1 + \frac{\varepsilon}{8} \quad (4.17)$$

$$D_1 H_{13} = \frac{\varepsilon[1 - \varepsilon/4(1/4 + v) - \varepsilon^2/8(437/480 - v)]}{12(1 - \varepsilon/8 - 19\varepsilon^2/960)} \quad (4.18)$$

$$D_1 H_{15} = -\frac{\varepsilon^2}{320(1 - \varepsilon/24)}[1 + \varepsilon/3(1/6 - v)] \quad (4.19)$$

$$D_1 H_{17} = \frac{\varepsilon^3}{14{,}336} \quad (4.20)$$

and

$$D_1 = 1 - \frac{\varepsilon}{2}\left(\frac{7}{4} - v\right) + \frac{3\varepsilon^2}{16}\left(\frac{7}{6} - v\right) - \frac{5\varepsilon^3}{96}\left(\frac{139}{120} - v\right) \quad (4.21)$$

It was first pointed out by Stewartson (Ref. 64), when treating the corresponding steady supersonic flow problem by use of Laplace transform, that it is necessary to take into account those poles of Φ_1 which are closest to the origin. At the inversion these lead in the present case to terms of the type $\exp(-i \text{ const. } x/\sigma^2 k)$, which may be numerically of the same order as terms like $(k\sigma^2)^n \ln(k\sigma^2)$.

The smallest zero of D_1 is found to be located at

$$\varepsilon = \varepsilon_0 = 0.192 + 0.381\,i \qquad (4.22)$$

which differs insignificantly from the results of Stewartson (Ref. 64) and Miles (Ref. 46).

There are, of course, also other poles but they seem, in accordance with Stewartson and Miles, to be located outside $|\varepsilon| = 4$, and the present method is anyhow incapable of determining them with any accuracy. Thus, for example, Eq. (4.18) indicates a pole at $\varepsilon = 4.6$, but this result is clearly spurious, since the physics of the problem make it evident that there cannot be any terms of the form $\cos(ax)$ or $\sin(ax)$ as $x \to \infty$ (which appear if the pole is located on the real axis).

Denoting by a_{11}, a_{13}, ... the residues of H_{11}, H_{13}, ... we find that

$$a_{11} = 0.088 + 0.486\,i \qquad (4.23)$$

$$a_{13} = 0.021 - 0.017\,i \qquad (4.24)$$

while a_{15} and a_{17} are negligible.

In order to simplify the inversion of the Fourier transform Eq. (4.16) we consider only the asymptotic solution for large values of x (or, actually, for large values of $x/\sigma^2 k$) which is determined from the behavior of the transform for small values of $|u|$. Therefore, terms which are non-analytic in this region are first separated out by setting

$$H_{11} = \frac{\varepsilon v}{2}\left[1 + \frac{\varepsilon}{2}\left(\frac{5}{2} - v\right)\right] + \frac{a_{11}}{\varepsilon - \varepsilon_0} + R_{11} \qquad (4.25)$$

$$H_{13} = -\frac{v\varepsilon^2}{16} + \frac{a_{13}}{\varepsilon - \varepsilon_0} + R_{13} \qquad (4.26)$$

Terms of order $\varepsilon^3 v$ and higher are now dropped henceforth. To order $\varepsilon^3 v$, R_{11}, and R_{13} may then be considered analytic in u and expanded in powers thereof, giving

$$R_{11} = -1 - \frac{3\varepsilon}{4} - \frac{7\varepsilon^2}{16} + \frac{2a_{11}}{k\sigma^2}\sum_{m=0}^{\infty}\left(\frac{2\varepsilon_0}{k\sigma^2} - \frac{k}{2}\right)^{-m-1} u^m \qquad (4.27)$$

$$R_{13} = \frac{\varepsilon}{12} + \frac{5\varepsilon^2}{64} + \frac{2a_{13}}{k\sigma^2}\sum_{m=0}^{\infty}\left(\frac{2\varepsilon_0}{k\sigma^2} - \frac{k}{2}\right)^{-m-1} u^m \qquad (4.28)$$

Terms of the form $u^p F(u)$, which now occur in the transform, give at the inversion $(-i\,\partial/\partial x)^p f(x)$ (asymptotically). In this way the asymptotic inversion formulae may be obtained in a simple way from ordinary inversion formulae. The asymptotic formulae are valid if the functions $w_n(x)$ are continuous and have continuous derivatives, as may be proved by expanding w_n in a power series in x.

The final result for $\varphi_1 = \mathfrak{F}^{-1}\{\Phi_1\}$ is then expressed as follows

$$\varphi_1 = \varphi_{11} + \varphi_{13} + \varphi_{15} + 0(k^3\sigma^6 \ln k\sigma^2) \tag{4.29}$$

where

$$
\begin{aligned}
\varphi_{11} = & -\left[1 - \frac{ik\sigma^2}{4}\left(\frac{3}{2} - v\right)D - \frac{k^2\sigma^4}{16}\left(\frac{7}{4} - \frac{5v}{2} - \frac{\pi^2}{6} + v^2\right)D^2\right]w_1(x) \\
& + \left[\frac{ik\sigma^2}{4}D^2 + \frac{k^2\sigma^4}{16}\left(\frac{5}{2} - 2v\right)D^3\right]\int_0^x \exp\left[-i\frac{k}{2}(x-\xi)\right]\ln(x-\xi) \\
& \times w_1(\xi)\,d\xi + \frac{k^2\sigma^4}{16}D^3\int_0^x \exp\left[-i\frac{k}{2}(x-\xi)\right]\ln^2(x-\xi)w_1(\xi)\,d\xi \\
& + a_{11}\left\{\frac{2i}{k\sigma^2}\int_0^x \exp\left[\frac{2i}{k\sigma^2}B(x-\xi)\right]w_1(\xi)\,d\xi \\
& + \sum_{m=0}^{\infty}B^{-m-1}\left(-i\frac{k\sigma^2}{2}\frac{\partial}{\partial x}\right)^m w_1(x)\right\}
\end{aligned}
\tag{4.30}
$$

$$
\begin{aligned}
\varphi_{13} = & -\left[\frac{ik\sigma^2}{32}\left(\frac{4}{3} - v\right)D + \frac{5k^2\sigma^4}{256}D^2\right]w_3(x) - \frac{ik\sigma^2}{24}D^2\int_0^x \exp\left[-i\frac{k}{2}(x-\xi)\right] \\
& \times \ln(x-\xi)w_3(\xi)\,d\xi + a_{13}\left\{\frac{2i}{k\sigma^2}\int_0^x \exp\left[\frac{2i}{k\sigma^2}B(x-\xi)\right]w_3(\xi)\,d\xi \\
& + \sum_{m=0}^{\infty}B^{-m-1}\left(-i\frac{k\sigma^2}{2}\frac{\partial}{\partial x}\right)^m w_3(x)\right\}
\end{aligned}
\tag{4.31}
$$

$$\varphi_{15} = \frac{k^2\sigma^4}{1280}D^2 w_5 \tag{4.32}$$

$$D = \frac{\partial}{\partial x} + \frac{ik}{2} \tag{4.33}$$

$$v = \gamma - \ln\frac{8}{k\sigma^2} + \frac{\pi i}{2} \tag{4.34}$$

$$B = \varepsilon_0 - \frac{k^2\sigma^2}{4} \tag{4.35}$$

Having Φ_1, we may now calculate Φ_3 in a like manner. For the calculation of poles and residues as for Φ_1, Φ_3 is actually needed to a higher approximation than that given by Eq. (4.15). However, it is soon found that the only pole to take into account is that of Φ_1 already determined, so that Eq. (4.15) is sufficient as it stands.

The final result for Φ_3 is then found to be

$$\Phi_3 = H_{31}W_1 + H_{33}W_3 + H_{35}W_5 + 0(\varepsilon^3 v) \tag{4.36}$$

where

$$H_{13} = H_{31} \tag{4.37}$$

$$H_{33} = R_{33} \tag{4.38}$$

$$H_{35} = R_{35} \tag{4.39}$$

$$R_{33} = \frac{1}{3}\left(1 + \frac{\varepsilon}{8} + \frac{9\varepsilon^2}{160}\right) + 0(\varepsilon^3) \tag{4.40}$$

$$R_{35} = \frac{\varepsilon}{120}\left(1 + \frac{\varepsilon}{4}\right) + 0(\varepsilon^3) \tag{4.41}$$

It is easily shown that, due to symmetry, $H_{nm} = H_{mn}$.

Upon inversion Eqs. (4.36)–(4.41) give

$$\varphi_3 = \varphi_{31} + \varphi_{33} + \varphi_{35} + 0(k^3\sigma^6 \ln k\sigma^2) \tag{4.42}$$

$$\varphi_{31} = -\left[\frac{ik\sigma^2}{32}\left(\frac{4}{3} - v\right)D + \frac{5k^2\sigma^4}{256}D^2\right]w_1(x) - \frac{ik\sigma^2}{24}D^2\int_0^x \exp\left[-i\frac{k}{2}(x-\xi)\right]$$

$$\times \ln(x-\xi)w_1(\xi)\,d\xi + a_{13}\left\{\frac{2i}{k\sigma^2}\int_0^x \exp\left[\frac{2i}{k\sigma^2}B(x-\xi)\right]w_1(\xi)\,d\xi\right.$$

$$\left. + \sum_{m=0}^\infty B^{-m-1}\left(-i\frac{k\sigma^2}{2}\frac{\partial}{\partial x}\right)^m w_1(x)\right\} \tag{4.43}$$

$$\varphi_{33} = -\frac{1}{3}\left(1 - \frac{ik\sigma^2}{16}D - \frac{9k^2\sigma^4}{1024}D^2\right)w_3(x) \tag{4.44}$$

$$\varphi_{35} = -\frac{1}{240}\left(ik\sigma^2 D + \frac{k^2\sigma^4}{8}D^2\right)w_5(x) \tag{4.45}$$

For $n \geq 5$ there are no poles to consider and the result may then be directly expanded in powers of ε leading to the following general formula

$$\varphi_n = \varphi_{n,n-4} + \varphi_{n,n-2} + \varphi_{n,n} + \varphi_{n,n+2} + \varphi_{n,n+4} \tag{4.46}$$

where

$$\varphi_{n,n-4} = \frac{3k^2\sigma^4}{32n(n-1)(n-2)(n-3)(n-4)}D^2\,w_{n-4} \tag{4.47}$$

$$\varphi_{n,n-2} = -\left[\frac{ik\sigma^2}{4n(n-1)(n-2)}D + \frac{3k^2\sigma^4}{8n(n^2-1)(n-2)(n-3)}D^2\right]w_{n-2} \tag{4.48}$$

$$\varphi_{n,n} = -\frac{1}{n}\left[1 - \frac{ik\sigma^2}{2(n^2-1)}D - \frac{9k^2\sigma^4}{16(n^2-1)(n^2-4)}D^2\right]w_n \tag{4.49}$$

$$\varphi_{n,n+2} = -\left[\frac{ik\sigma^2}{4n(n+1)(n+2)}D + \frac{3k^2\sigma^4}{8n(n^2-1)(n+2)(n+3)}D^2\right]w_{n+2} \tag{4.50}$$

$$\varphi_{n,n+4} = \frac{3k^2\sigma^4}{32n(n+1)(n+2)(n+3)(n+4)}D^2 w_{n+4} \tag{4.51}$$

As n increases φ_n thus tends to its slender-wing-theory value $-(1/n)w_n$.

To complete the analysis we notice that Eq. (4.9) defines the terms in a Fourier sine series (a Glauert series), which immediately gives

$$\varphi(x, \psi) = \frac{2\sigma}{\pi} \sum_{n=1}^{\infty} \varphi_n(x) \sin n\psi \qquad (4.52)$$

The problem is now solved for any distribution of $w(x, y)$ which is symmetric in y. For antisymmetric upwash distributions one obtains a similar set of equations for even values of n which could be solved in the same manner.

It is noted that, with the exception of the exponential terms, the result given above could also have been obtained by use of the Adams–Sears iteration scheme employed in Chapter 3 if the original w-distribution were modified by a smoothing process so that singularities did not occur in the higher order terms. The exponential terms give the asymptotic representation for large x of the reflections of the strong leading-edge disturbance by the side edges. Jumps of w or its derivatives in the x-direction will similarly lead to exponential terms originating at the points of discontinuities, as can be seen by integrating the exponential terms in Eqs. (4.30), (4.31) and (4.43) by parts. For the low aspect ratio theory to be valid the jumps must not occur too close to the trailing edge (Ref. 46). Further developments are considered in Sections 4.4 and 8.5.

4.3. Generalized aerodynamic forces

When introducing the results for φ given above into Eqs. (1.60)–(1.64) for the calculation of aerodynamic forces it is seen that, in general, the singularity at the leading edge will be non-integrable. This difficulty may be overcome as follows. Assuming that the mode functions f_j are polynomials, Eqs. (1.60) and (1.62) lead to integrals of the type

$$\int_0^1 x^q \varphi_x \, dx \quad \text{and} \quad \int_0^1 x^q \, \varphi \, dx \qquad (4.53)$$

Through a number of integration by parts these may be written as a combination of integrals like

$$\int_0^x \int_0^x \dots \int_0^1 \varphi(dx)^q \qquad (4.54)$$

Considering these as functions of the upper limit x their transforms are simply $(iu)^{-q}\Phi$ and their asymptotic expansions for large x may be obtained in the same manner as that for φ and hence the weighted integral in Eq. (1.62) to the same degree of approximation as φ. This is essentially the method used by Miles (Ref. 45). Now it is actually not necessary to go back as far as to the transformed potential, but the multiple integral may be obtained directly from the result for φ by substituting $\int_0^x \int_0^x \dots \int_0^x w(x, y)(dx)^q$ for $w(x, y)$. Thus L_{11}, for example, may be obtained simply as $(2/\sigma)\int_{-\sigma}^{\sigma}\varphi_{L_{11}}(1,y)\,dy$ where $\varphi_{L_{11}}$ is the potential resulting from setting $w = ik(1 + ikx)$. In this manner all force coefficients can be obtained. Similarly as in Chapter 3 the final results for the total-force coefficients are most conveniently expressed in the following way:

$$L_{ij} = L_{ij}^{(1)} + L_{ij}^{(2)} + L_{ij}^{(3)} + \bar{E}_{ij} \qquad (4.55)$$

where $L_{ij}^{(1)}$ represents the slender-wing solution, $L_{ij}^{(2)}$ is the term of relative order $k\sigma^2$ and $L_{ij}^{(3)}$ that of order $k^2\sigma^4$. These terms are similar to those of Chapter 3 and could, as a matter of fact, be obtained by applying the Adams–Sears method and thereby discount infinite terms in integration limits whenever such occur. The exponential terms $\bar{E}_{ij}$ did not appear for delta-type wings. Restricting the result for low k as in Chapter 3 gives

$$\frac{2}{\pi A} L_{11}^{(1)} = -ik + k^2 \qquad (4.56)$$

$$\frac{2}{\pi A} L_{11}^{(2)} = -\frac{k^2\sigma^2}{4}\left[1 - \frac{3ik}{2}v + \frac{7}{4}ik + 0(k^2)\right] \qquad (4.57)$$

$$\frac{2}{\pi A} L_{11}^{(3)} = \frac{ik^3\sigma^4}{16}\left[2v - \frac{1}{2} + 0(k)\right] = ik\left[\frac{2}{\pi A} L_{21}^{(3)}\right] \qquad (4.58)$$

$$\frac{2}{\pi A} L_{12}^{(1)} = \frac{k^2}{2} \qquad (4.59)$$

$$\frac{2}{\pi A} L_{12}^{(2)} = -\frac{k^2\sigma^2}{4}\left[v - \frac{1}{2} + \frac{3ik}{2} + 0(k^2)\right] \qquad (4.60)$$

$$\frac{2}{\pi A} L_{12}^{(3)} = \frac{ik^3\sigma^4}{16}\left[4v - 3 + 0(k)\right] = ik\left[\frac{2}{\pi A}L_{22}^{(3)}\right] \qquad (4.61)$$

$$\frac{2}{\pi A} L_{21}^{(1)} = -1 - 2ik + \frac{k^2}{2} \qquad (4.62)$$

$$\frac{2}{\pi A} L_{21}^{(2)} = \frac{ik\sigma^2}{4}\left[1 - \frac{5ik}{2}v + \frac{13}{4}ik + 0(k^2)\right] \qquad (4.63)$$

$$\frac{2}{\pi A} L_{22}^{(1)} = -ik + \frac{k^2}{3} \qquad (4.64)$$

$$\frac{2}{\pi A} L_{22}^{(2)} = \frac{ik\sigma^2}{4}\left[v - \frac{1}{2} + \frac{5}{2}ik + 0(k^2)\right] \qquad (4.65)$$

and

$$\bar{E}_{11} = \frac{ika_{11}}{B}\left(1 + \frac{k^2\sigma^2}{2B}\right)e^{2iB/k\sigma^2} \qquad (4.66)$$

$$\bar{E}_{12} = \left(1 + \frac{ik\sigma^2}{2B}\right)\bar{E}_{11} \qquad (4.67)$$

$$\bar{E}_{21} = \frac{a_{11}}{B}\left(1 + \frac{k^2\sigma^2}{2B}\right)e^{2iB/k\sigma^2} \qquad (4.68)$$

$$\bar{E}_{22} = \left(1 + \frac{ik\sigma^2}{2B}\right)\bar{E}_{21} \qquad (4.69)$$

In Figs. 4.1 and 4.2 are shown some results for $L_{11} = |L_{11}| \exp(i\theta_{11})$ according to the different order approximations. For the $A = 1$ wing the second- and third-order solutions, as well as the complete solution (with the exponential

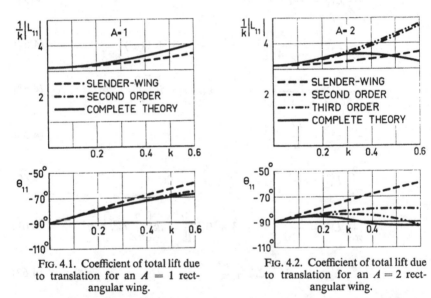

FIG. 4.1. Coefficient of total lift due to translation for an $A = 1$ rectangular wing.

FIG. 4.2. Coefficient of total lift due to translation for an $A = 2$ rectangular wing.

term included) give practically identical results for the range of reduced frequency treated. The third-order and exponential terms become of importance only for frequencies higher than those shown. Actually, the error of the slender-wing solution is in this case only 10 per cent for the force magnitude and 8.3° for the phase angle at $k = 0.6$. For the $A = 2$ wing, on the other hand, the higher order terms are all important, even for fairly low k. The exponential term turns out to be somewhat larger in magnitude than the third-order term.

Further numerical results are given in Chapters 6 and 7.

4.4. Stability derivatives for a cropped delta wing

The planform considered is shown in Fig. 4.3.

The Adams–Sears iteration method described in Chapter 3 could not be used for this wing since the leading edge slope is discontinuous at $x = x_1$.

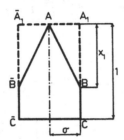

FIG. 4.3. Cropped delta wing.

However, for $x < x_1$ we have an ordinary delta wing and for this part φ can be found from Eqs. (3.10)–(3.17). Hence, for w independent of y and low frequencies we find that

$$\varphi = -\sqrt{(s^2 - y^2)} \left\{ w + \frac{ik}{4} \frac{\partial}{\partial x} \left[\left(v + 2 \ln \frac{s}{\sigma} - 2 \right) ws^2 \right. \right.$$
$$\left. \left. - \frac{\partial}{\partial x} \int_0^x \ln (x - \xi) ws^2 d\xi \right] - \frac{ik}{6} \left(\frac{s^2}{2} + y^2 \right) w_x \right\} \tag{4.70}$$

For the calculation of φ for $x > x_1$ we consider the problem for the rectangular wing $A_1 C \bar{C} \bar{A}_1$ shown in Fig. 4.3. This wing can be treated by the method of Section 4.2 if we consider w known in the regions AA_1B and $\bar{A}\bar{B}\bar{A}_1$. There w must be chosen so as to make φ zero outside the original wing. For $x < x_1$ the result obtained in this way will be that given by Adams–Sears' method, Eq. (4.70). For $x > x_1$, however, there will also appear the exponential terms associated with the jumps in the w_n:s or their x-derivatives at $x = x_1$. In order to calculate total forces and moments we need the spanwise integral of φ, i.e. φ_1. The final result for φ_1 may be written

$$\varphi_1 = (\varphi_1)_{\text{Adams–Sears}} + E_{11} + E_{13} \tag{4.71}$$

where E_{11} and E_{13} are the exponential terms given by the curly brackets in Eqs. (4.30) and (4.31), respectively. Through integration by parts and neglecting higher order terms in k these may be written

$$E_{11} = \frac{a_{11}}{B} \left[\Delta w_1(x_1) + \frac{k\sigma^2}{2iB} \Delta w_{1x}(x_1) \right] \exp \left[\frac{2iB}{k\sigma^2} (x - x_1) \right] \tag{4.72}$$

$$E_{13} = \frac{a_{13}}{B} \left[\Delta w_3(x_1) + \frac{k\sigma^2}{2iB} \Delta w_{3x}(x_1) \right] \exp \left[\frac{2iB}{k\sigma^2} (x - x_1) \right] \tag{4.73}$$

where $\Delta w_n(x_1)$ is the jump in w_n at $x = x_1$ and $\Delta w_{nx}(x_1)$ the jump in w_{nx}. For $x > x_1$, $w_1 = (\pi/2)w$ and $w_3 = 0$. For $x < x_1$ it is possible to express w_1 in terms of φ_1 and φ_3 by aid of Eq. (4.10). After inversion and neglecting higher order terms in k, we find that

$$w_1 = -\varphi_1 + \frac{ik\sigma^2}{4}\left(v - \frac{3}{2}\right)\varphi_{1x} - \frac{ik\sigma^2}{4}\frac{\partial^2}{\partial x^2}\int_0^x \ln(x - \xi)\varphi_1(\xi)\,d\xi + \frac{ik\sigma^2}{8}\varphi_{3x} \tag{4.74}$$

Now

$$\varphi_1 = \frac{1}{\sigma}\int_0^\pi \varphi \sin\theta\,d\theta = \frac{1}{\sigma^2}\int_{-s}^s \varphi\,dy \tag{4.75}$$

and

$$\varphi_3 = \frac{1}{\sigma}\int_0^\pi \varphi \sin 3\theta\,d\theta = \frac{1}{\sigma^2}\int_{-s}^s \left(4\frac{y^2}{\sigma^2} - 1\right)\varphi\,dy \tag{4.76}$$

Using Eq. (4.70) φ_1 and φ_3 may be obtained and then Δw_1 and Δw_{1x} calculated. Neglecting terms of order $k\sigma^2$ and higher we find that

$$w_1 = \frac{\pi s^2}{2\sigma^2}w \tag{4.77}$$

Hence $\Delta w_1 = 0$ and

$$\Delta w_{1x} = -\left[\frac{\pi s s_x}{\sigma^2}w\right]_{x=x_1} = -\pi \frac{w(x_1)}{x_1} \tag{4.78}$$

It turns out that the second-order terms do not give any contribution to Δw_1. Hence

$$E_{11} = -\frac{a_{11}}{B}\pi \frac{w(x_1)}{x_1}\frac{k\sigma^2}{2iB}\exp\left[\frac{2iB}{k\sigma^2}(x - x_1)\right] + 0(k^2) \tag{4.79}$$

Similarly, inverting Eq. (4.14) gives for w_3

$$w_3 = \frac{ik\sigma^2}{8}\varphi_{1x} - 3\left(1 + \frac{ik\sigma^2}{16}\frac{\partial}{\partial x}\right)\varphi_3 + 0(k^2) \tag{4.80}$$

φ_1 and φ_3 are obtained as before from Eqs. (4.75), (4.76) and (4.70). Upon substitution in Eq. (4.80) and some calculations we find that

$$\Delta w_3 = -\frac{\pi i k\sigma^2}{16}\frac{w(x_1)}{x_1} + 0(k^2) \tag{4.81}$$

$$\Delta w_{3x} = -3\pi \frac{w(x_1)}{x_1} + 0(k) \tag{4.82}$$

Thus

$$E_{13} = -\frac{a_{13}}{B}\pi \frac{w(x_1)}{x_1}\left(3\frac{k\sigma^2}{2iB} + \frac{ik\sigma^2}{16}\right)\exp\left[\frac{2iB}{k\sigma^2}(x - x_1)\right] \tag{4.83}$$

It is now possible to calculate the stability derivatives. The unsteady derivatives are given by Eqs. (3.42) and (3.43) plus the exponential terms. By carrying out the integrations we find that

$$\frac{2}{\pi A}(C_{Lq} + C_{L\dot{\alpha}}) = 2 - a - \frac{2}{3}x_1 - \frac{\sigma^2}{2x_1}\left(\frac{1}{x_1}\ln\frac{1}{1-x_1} + \bar{E}_c' - 1\right) \quad (4.84)$$

$$\frac{2}{\pi A}(C_{Mq} + C_{M\dot{\alpha}}) = -(1-a)^2 - \frac{\sigma^2}{8}\left[\bar{v} + f(x_1) - \frac{4}{x_1}(1-a)\right.$$

$$\left. \times \left(\frac{1}{x_1}\ln\frac{1}{1-x_1} + \bar{E}_c' - 1\right)\right] \quad (4.85)$$

where

$$\bar{E}_c' = 2\mathrm{Re}\left\{B^{-2}\left(a_{11} + 3a_{13} + \frac{1}{8}Ba_{13}\right)\exp\left[\frac{2iB}{k\sigma^2}(1-x_1)\right]\right\} \quad (4.86)$$

$$\bar{v} = \ln\frac{8}{k\sigma^2} - \gamma$$

and

$$f(x_1) = 9/4 - 2x_1^{-1} - \ln x_1 - 2(x_1^{-2} - 1)\ln(1-x_1) \quad (4.87)$$

Introducing the numerical values of a_{11}, a_{13} and B from Section 4.2 gives

$$\bar{E}_c' = e^{-\beta}(2.86\cos\alpha + 4.19\sin\alpha) \quad (4.88)$$

where

$$\beta = \frac{0.762}{k\sigma^2}(1-x_1) \quad (4.89)$$

$$\alpha = \frac{0.382}{k\sigma^2}(1-x_1) \quad (4.90)$$

It may also be of some interest to calculate the first-order frequency effect on the quasi-steady derivatives. These are given by

$$\frac{2}{\pi A}C_{L\alpha} = 1 + \frac{k\sigma^2}{2x_1}\bar{E}_c'' \quad (4.91)$$

$$\frac{2}{\pi A}C_{M\alpha} = -\frac{2x_1}{3} + a - \frac{k\sigma^2}{8}\left[\frac{\pi}{2} - \frac{4}{x_1}(1-a)\bar{E}_c''\right] \quad (4.92)$$

where

$$\bar{E}_c'' = 2\mathrm{Im}\left\{B^{-2}\left(a_{11} + 3a_{13} + \frac{1}{8}Ba_{13}\right)\exp\left[\frac{2iB}{k\sigma^2}(1-x_1)\right]\right\}$$

$$= (4.19\cos\alpha - 2.86\sin\alpha)e^{-\beta} \quad (4.93)$$

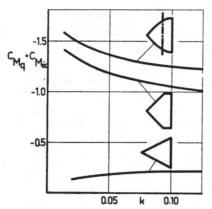

FIG. 4.4. Damping in pitch for a cropped delta wing of $A = 2$. $a = 0.6$.

In Fig. 4.4 is plotted $C_{Mq} + C_{M\alpha}$ for a cropped delta with $A = 2$ and $x_1 = 2/3$
together with results for the pure delta and for the convex wing of Section 3.5
all with pitching axis at $a = 0.6$. Apparently a rounded planform is not necessary
in order to achieve the favorable effect on damping of cutting off the wing tips.

CHAPTER 5

LOW ASPECT RATIO WING–BODY COMBINATIONS

5.1. Introduction

As was shown in Chapter 1 non-linear thickness effects can be neglected when the requirement (1.40) is fulfilled. However, for non-planar problems there may still be thickness effects because of the tangency condition being applied on the true boundary of the configuration and not on $z = 0$. These may be computed from linearized theory if the frequency is sufficiently high. The method described in the present chapter for the calculation of dynamic stability derivatives of low aspect ratio wing–body combinations (Ref. 31) is an extension of the iteration method of Chapter 3.

5.2. Method of solution for arbitrary cross-sectional shape

The problem is to find a solution of Eq. (1.88) subject to the tangency condition,

$$\varphi_n = \delta(f_x + ikf) \sin v \tag{5.1}$$

where n is the outward normal making the angle v with the y-axis (see Fig. 1.1). Eq. (5.1) may be applied at the mean position of the surface of the configuration since only very small oscillations are studied.

If the cross-section is sufficiently flat the solution can be built up from a distribution of doublets on the x, y-plane. An approximate doublet distribution solution valid for shapes of small span is given, in the Fourier transformed plane, by Eq. (3.4). Inverting to the physical plane and retaining only first- and second-order terms gives

$$\varphi(x, y, z) = \frac{1}{\pi} \int_{-s}^{s} \frac{z\varphi}{\tilde{r}^2} \, d\eta + \frac{ik}{2\pi} z D \left\{ \left(\ln \frac{k}{2} + \gamma + \frac{\pi i}{2} - 1 \right) \int_{-s}^{s} \varphi \, d\eta \right.$$

$$\left. + 2 \int_{-s}^{s} \varphi \ln \tilde{r} \, d\eta - D \int_{0}^{x} \exp \left[-i\frac{k}{2}(x - \xi) \right] \ln (x - \xi) \, d\xi \left[\int_{-s}^{s} \varphi \, d\eta \right] \right\} \tag{5.2}$$

where

$$\tilde{r} = \sqrt{[(y - \eta)^2 + z^2]}$$

and

$$D = \frac{\partial}{\partial x} + i\frac{k}{2}$$

(For shortness the argument of $\varphi(x, \eta, +0)$ is left out in the right-hand side of Eq. (5.2).) The problem is now to find a doublet distribution such that the tangency condition, Eq. (5.1), is fulfilled. By extending the method of Chapter 3 an approximate solution can be given as follows:

$$\varphi = \varphi^{(1)} + \varphi^{(2)} \tag{5.3}$$

where

$$\varphi^{(1)} = \frac{1}{\pi} \int_{-s}^{s} \frac{z\varphi^{(1)}(x, \eta, +0)\, d\eta}{\tilde{r}^2} \tag{5.4}$$

$$\varphi^{(2)} = \psi^{(2)} + F^{(1)} \tag{5.5}$$

$$\psi^{(2)} = \frac{1}{\pi} \int_{-s}^{s} \frac{z\psi^{(2)}(x, \eta, +0)\, d\eta}{\tilde{r}^2} \tag{5.6}$$

and

$$F^{(1)} = \frac{ik}{2\pi} zD \left\{ \left(\ln \frac{k}{2} + \gamma + \frac{\pi i}{2} - 1 \right) \int_{-s}^{s} \varphi^{(1)}(x, \eta, +0)\, d\eta \right.$$

$$+ 2 \int_{-s}^{s} \varphi^{(1)}(x, \eta, +0) \ln \tilde{r}\, d\eta - D \int_{0}^{x} \exp \left[-i \frac{k}{2} (x - \xi) \right]$$

$$\left. \times \ln (x - \xi)\, d\xi \left[\int_{-s}^{s} \varphi^{(1)}(\xi, \eta, +0)\, d\eta \right] \right. \tag{5.7}$$

$\varphi^{(1)}$ and $\psi^{(2)}$ are thus solutions of the two-dimensional Laplace equation in each cross-section $x = $ constant. $\varphi^{(0)}$ is first determined so that the tangency condition is fulfilled. Hence we obtain the usual slender-body result (Ref. 68). For the calculation of $\varphi^{(1)}$ we introduce the proper conformal transformation that maps the cross-section of the configuration onto a slit along the real axis. Let this transformation be given by

$$X_1 = T(X); \qquad X = \tau(X_1) \tag{5.8}$$

where†

$$X = y + jz; \qquad X_1 = y_1 + jz_1 \tag{5.9}$$

In the transformed plane the outside of the cross-section is mapped onto the outside of a slit along the y_1-axis for $|y_1| < s_1$ as shown in Fig. 5.1. The transformation is so chosen that both axes retain their positions and infinity remains undistorted.

The tangency condition is transformed to the X_1-plane by means of the formula

$$\varphi_{y_1}^{(1)} - j\varphi_{z_1}^{(1)} = (\varphi_y^{(1)} - j\varphi_z^{(1)}) \frac{d\tau}{dX_1} \tag{5.10}$$

---

† The symbol j is used for the imaginary unit during conformal transformation to distinguish it from i used in connection with harmonic oscillations.

When inserted in Eq. (5.1) this will prescribe a value of $\varphi_{z_1}^{(1)}$ along the y_1-axis for $|y_1| < s_1$. The problem of finding the proper doublet distribution along the slit giving this value of $\varphi_z^{(1)}$ leads to the integral equation of thin airfoil theory

$$\varphi_{z_1}^{(1)}(y_1, 0) = -\frac{1}{\pi} \oint_{-s}^{s} \frac{\varphi_{\eta_1}^{(1)}(\eta_1, +0)\, d\eta_1}{y_1 - \eta_1} \tag{5.11}$$

The solution having $\varphi(\pm s_1, 0) = 0$ is given by the well-known inversion formula (Ref. 61):

$$\varphi_{y_1}^{(1)} = \frac{1}{\pi\sqrt{(s_1{}^2 - y_1{}^2)}} \int_{-s_1}^{s_1} \frac{\varphi_{z_1}^{(1)}(\eta_1, 0)\sqrt{(s_1{}^2 - \eta_1{}^2)}\, d\eta_1}{y_1 - \eta_1} \tag{5.12}$$

By means of Eq. (5.8) the result can then be transformed back to the original plane.

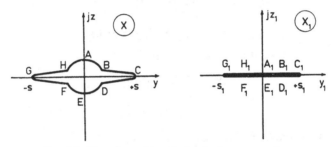

FIG. 5.1. Transformation of wing-body cross-section.

To get $\varphi^{(2)}$, the result thus obtained for $\varphi^{(1)}$ is inserted in Eq. (5.7) and $F^{(1)}$ calculated. Since $\psi^{(2)}$, also, is a solution of the two-dimensional Laplace equation in the cross-flow plane it can be obtained in the same way as $\varphi^{(1)}$. The boundary condition on the body surface for $\psi^{(2)}$ follows from Eq. (5.5). Now Eq. (5.1) is already fulfilled by $\varphi^{(1)}$ so that we must have

$$\psi_n^{(2)} = -F_n^{(1)} \tag{5.13}$$

From here on the analysis for $\psi^{(2)}$ is identical to that for $\varphi^{(1)}$. The final second-order solution is then given by

$$\varphi = \varphi^{(1)} + \varphi^{(2)} + F^{(1)} + 0[(k\sigma^2 \ln k\sigma^2)^2] \tag{5.14}$$

5.3. Solution for a slowly oscillating winged body of revolution

A slowly oscillating combination of a wing of zero thickness mounted centrally on a body of revolution will be treated. A cross-section of the configuration is shown in Fig. 5.2. For application of the tangency condition, the direction of the normal on the body is the r-direction in the cylindrical co-ordinate system x, r, θ.

Calculation of stability derivatives requires solutions of φ for the modes f_1 d f_2 given in Section 1.8. However, if only slow oscillations are considered it

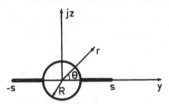

FIG. 5.2. Cross-section of winged body of revolution.

turns out that the required solutions can be constructed from the basic solution φ_0, defined by the boundary conditions

$$\varphi_{0r} = \sin\theta \quad \text{on the body} \tag{5.15}$$

and

$$\varphi_{0z} = 1 \quad \text{on the wing } (z = 0) \tag{5.16}$$

The proper conformal transformation is

$$X_1 = \tfrac{1}{2}(X + R^2/X) \tag{5.17}$$

Hence

$$s_1 = \tfrac{1}{2}(s + R^2/s) \tag{5.18}$$

Solving for X in terms of X_1 gives

$$X = \tau(X_1) = X_1 + (X_1{}^2 - R^2)^{1/2} \tag{5.19}$$

The correct branch of the square root is obtained by inserting a cut in the X_1-plane between the points $X_1 = \pm s_1$.

According to Eq. (5.10) velocities in the two planes are related through

$$\varphi_{0y_1}^{(1)} - j\varphi_{0z_1}^{(1)} = (\varphi_{0y}^{(1)} - j\varphi_{0z}^{(1)})\left[1 + \frac{X_1}{(X_1{}^2 - R^2)^{1/2}}\right] \tag{5.20}$$

Insertion of Eqs. (5.15), (5.16) and (5.19) then gives

$$\varphi_{0z_1}^{(1)} = 1 \quad \text{for } |y_1| < R \tag{5.21}$$

$$\varphi_{0z_1}^{(1)} = 1 + \frac{|y_1|}{\sqrt{(y_1{}^2 - R^2)}} \quad \text{for } R < |y_1| < s_1 \tag{5.22}$$

By introducing Eqs. (5.21) and (5.22) into Eq. (5.12) $\varphi_0^{(1)}$ is then found. The integral is most easily found by contour integration in the complex X_1-plane. As shown in Ref. 31 the result becomes

$$\varphi_0^{(1)}(y_1, +0) = -2\sqrt{(s_1{}^2 - y_1{}^2)} + \operatorname{Re}\{\sqrt{R^2 - y_1{}^2}\} \tag{5.23}$$

Analytic continuation in the X_1-plane gives, since $\varphi_0^{(1)}$ should vanish at infinity,

$$\varphi_0^{(1)} = \mathrm{Re}\,\{-2(s_1{}^2 - X_1{}^2)^{1/2} + (R^2 - X_1)^{1/2} - jX_1\} \qquad (5.24)$$

In order to select the proper branches of the square roots the cut between $X_1 = \pm s_1$ is utilized.

Returning to the original plane gives the slender-body solution (Ref. 68)

$$\varphi_0^{(1)} = -\mathrm{Re}\,\{[(s + R^2/s)^2 - (X + R^2/X)^2]^{1/2} + jX\} \qquad (5.25)$$

To render Eq. (5.25) single-valued we introduce two cuts along the real axis between the branch-points $y = s, y = R^2/s$ and $y = -s, y = -R^2/s$ respectively. Eq. (5.25) shows that the first-order singularity distribution consists of a continuous doublet distribution for $R^2/s \leq |y| \leq s$ and a concentrated doublet at the origin.

To calculate $F_0^{(1)}$ we need the following two integrals calculated in Ref. 31 by contour integration

$$\frac{1}{\pi} \int_{-s}^{s} \varphi_0^{(1)}(\eta, +0)\, d\eta = -\tfrac{1}{2}(s^2 + R^4/s^2) \qquad (5.26)$$

and

$$\frac{1}{\pi} \int_{-s}^{s} \varphi_0^{(1)}(\eta, +0) \ln \tilde{r}\, d\eta = -\tfrac{1}{2}\mathrm{Re}\,\Big\{ X^2 - XY + \tfrac{1}{2}(s + R^2/s)^2 \cosh^{-1} \frac{X + R^2/X}{s + R^2/s}$$
$$+ \tfrac{1}{2}(s - R^2/s)^2 \ln \frac{Y - j(X - R^2/X)}{s - R^2/s} + f_0 \Big\} \qquad (5.27)$$

where

$$Y = [(X + R^2/X)^2 - (s + R^2/s)^2]^{1/2}$$

and

$$f_0 = -\tfrac{1}{2}(s^2 + R^4/s^2)(1 + \ln 4) + \tfrac{1}{2}(s + R^2/s)^2 \ln (s + R^2/s)$$
$$+ \tfrac{1}{2}(s - R^2/s)^2 \ln (s - R^2/s)$$

From these results we get on the wing

$$F_0^{(1)} = 0$$

and on the body (upper surface)

$$F_0^{(1)} = -\tfrac{1}{2}ikR \sin\theta \left[\left(\frac{1}{2}\ln \frac{Y_b + 2R \sin\theta}{s - R^2/s}\right) \frac{\partial}{\partial x}(s - R^2/s)^2 + G_0 \right] \qquad (5.28)$$

where

$$Y_b = \sqrt{[(s + R^2/s)^2 - 4R^2 \cos^2\theta]} \qquad (5.29)$$

and

$$G_0 = f_{0x} + \frac{1}{2}\frac{\partial}{\partial x}\left[\left(\ln \frac{k}{2} + \gamma + \frac{\pi i}{2} - 1\right)\left(s^2 + \frac{R^4}{s^2}\right) \right.$$
$$\left. - \frac{\partial}{\partial x}\int_0^x \ln (x - \xi)\left(s^2 + \frac{R^4}{s^2}\right) d\xi \right] \qquad (5.30)$$

In $F_0^{(1)}$ we have neglected all higher order terms in k. For the calculation of

$F_{0n}^{(1)}$ from Eq. (5.10) we need $F_{0y}^{(1)}$ and $F_{0z}^{(1)}$ on the contour. After some calculations we get

$$F_{0z}^{(1)} = -\tfrac{1}{2}ikG_0 \quad \text{on the wing} \tag{5.31}$$

and

$$F_{0y}^{(1)} = -4ik\,Y_b^{-1}R^2R_x \cos\theta \sin^2\theta \tag{5.32}$$

$$F_{0z}^{(1)} = \frac{1}{2}\,ik\left\{ Y_b^{-1}\left[8R^2R_x \cos^2\theta \sin\theta - R\sin\theta\,\frac{\partial}{\partial x}(s + R^2/s)^2 \right]\right.$$

$$\left. -\frac{1}{2}\left(\ln\frac{Y_b}{s - R^2/s}\right)\frac{\partial}{\partial x}(s - R^2/s)^2 - G_0 \right\} \tag{5.33}$$

on the body.

These results are then combined by means of Eq. (5.10) to give $F_{0z_1}^{(1)}$ in the transformed plane. The result of this operation can be written

$$F_{0z_1}^{(1)}(y_1, 0) = -\frac{1}{2}\,ikG_0\text{Re}\left\{1 + \frac{|y_1|}{\sqrt{(y_1{}^2 - R^2)}}\right\}$$

$$-\frac{1}{4}\,ik\,\text{Re}\left\{\left[\ln\frac{2[\sqrt{(s_1{}^2 - y_1{}^2)} + \sqrt{(R^2 - y_1{}^2)}]}{s - R^2/s}\right.\right.$$

$$\left.\left.\times\frac{\partial}{\partial x}(s - R^2/s)^2 + \sqrt{\left(\frac{R^2 - y_1{}^2}{s_1{}^2 - y_1{}^2}\right)}\frac{\partial}{\partial x}(s + R^2/s)^2\right\} \tag{5.34}\right.$$

The boundary condition for $\psi_0^{(2)}$ along the y_1-axis is, according to Eq. (5.13),

$$\psi_{0z_1} = -F_{0z_1}^{(1)}(y_1, 0) \quad \text{for } |y_1| < s_1 \tag{5.35}$$

Application of the inversion formula, Eq. (5.12), then gives, using the integral formulas given in Ref. 31,

$$\psi_{0y_1}^{(2)} = \frac{1}{2}\,ikG_0\left[\frac{2y_1}{\sqrt{(s_1{}^2 - y_1{}^2)}} - \text{Re}\left\{\frac{y_1}{\sqrt{(R^2 - y_1{}^2)}}\right\}\right]$$

$$-\frac{1}{4}\,ik\left[\text{sgn}\,y_1\,\text{Re}\left\{\tan^{-1}\sqrt{\left(\frac{y_1{}^2 - R^2}{s_1{}^2 - y_1{}^2}\right)}\right\} + \tan^{-1}\frac{\sqrt{(s_1{}^2 - y_1{}^2)}}{|y_1|} - \frac{\pi}{2}\right\}$$

$$-\frac{y_1}{\sqrt{(s_1{}^2 - y_1{}^2)}}\ln\frac{s + R^2/s}{s - R^2/s}\right]\frac{\partial}{\partial x}(s - R^2/s)^2 + \frac{1}{4}\,ik\left[\frac{y_1}{\sqrt{(s_1{}^2 - y_1{}^2)}}\right.$$

$$\left. - \text{Re}\left\{\text{sgn}\,y_1\sqrt{\left(\frac{y_1{}^2 - R^2}{s_1{}^2 - y_1{}^2}\right)}\right\}\right]\frac{\partial}{\partial x}(s + R^2/s)^2 \tag{5.36}$$

By a second integration $\psi_0^{(2)}$ on the y_1-axis can be obtained and then, from Eq. (5.17), its value on the contour. However, we are primarily interested in total forces and moments for which, according to Eqs. (1.78) and (1.79), the spanwise integral of $\Delta\varphi$ is needed. Then, for a harmonic potential ψ, it can be shown by contour integration that

$$\int_{-s}^{s}\Delta\psi\,dy = -\int_{-s_1}^{s_1}\Delta\psi_{y_1}\,\text{Re}\,[y_1 + \text{sgn}\,y_1\,\sqrt{(y_1{}^2 - R^2)}]\,dy_1 \tag{5.37}$$

The full expression for $\int_{-s}^{s} \Delta\varphi_0^{(2)}\, dy$ as given in Ref. 31 is somewhat lengthy and contains terms like those in the expression for G_0, Eq. (5.30). Considerable simplifications result, however, from expanding the expression for small values of R/s which gives

$$\int_{-s}^{s} \Delta\varphi_0^{(2)}\, dy = -\frac{\pi}{4}\, ik\, \frac{\partial}{\partial x}\left[\left(\ln\frac{ks^2}{8} + \gamma + \frac{\pi i}{2} - 2\right)s^2\right.$$
$$\left. - \frac{\partial}{\partial x}\int_0^x \ln(x-\xi)s^2\, d\xi\right] - 2\pi i k R^2 s s_x + \frac{8}{3} ik R^3 s_x + 0(R^4)$$
$$(5.38)$$

5.4. Stability derivatives for a configuration with small body-radius-to-wing-span ratio

Most practical airplane or missile configurations are characterized by a fairly small body radius—say, that the maximum radius is less than 0.3 of the wing semi-span—so that Eq. (5.38) is a valid approximation. Application of the formulas in Section 1.8 for the calculation of stability derivatives shows that for the quasi-steady derivatives the body contribution is properly given by slender-body theory, namely

$$C_{L\alpha} = C_{L\alpha}^{(w)} - \frac{\pi A}{2}(\bar{R}^2 - \bar{R}^4) \tag{5.39}$$

$$C_{M\alpha} = C_{M\alpha}^{(w)} + \frac{\pi A}{2}\left[(1-a)(\bar{R}^2 - \bar{R}^4) - \sigma^{-2}\int_0^1 (R^2 - R^4/s^2)\, dx\right] \tag{5.40}$$

where $\bar{R}$ is the value of R/s for $x=1$ and $C_{L\alpha}^{(w)}$ and $C_{M\alpha}^{(w)}$ are the derivatives for the wing without body (where the wing planform has been suitably extended inside the body). These are given in Chapter 3. For the unsteady derivatives the results become

$$C_{Lq} + C_{L\dot\alpha} = C_{Lq}^{(w)} + C_{L\dot\alpha}^{(w)} - \frac{\pi A}{2}\left[(1-a)(\bar{R}^2 - \bar{R}^4)\right.$$
$$\left. + \sigma^{-2}\int_0^1 (R^2 - R^4/s^2)\, dx - \left(2\bar{R}^2 - \frac{8}{3\pi}\bar{R}^3\right)\sigma s_x(1)\right] \tag{5.41}$$

$$C_{Mq} + C_{M\dot\alpha} = C_{Mq}^{(w)} + C_{M\dot\alpha}^{(w)} + \frac{\pi A}{2}\left[(1-a)^2(\bar{R}^2 - \bar{R}^4)\right.$$
$$- (1-a)\left(2\bar{R}^2 - \frac{8}{3\pi}\bar{R}^3\right)\sigma s_x(1)$$
$$\left. + \sigma^{-2}\int_0^1\left(2R^2 s - \frac{8}{3\pi}R^3\right)s_x\, dx\right] \tag{5.42}$$

The results were obtained by assuming that $(R/s)^4$ was small compared to unity everywhere. Near the section where the leading edge intersects the body this is obviously not so and the expansion (5.38) not valid. However, for sections where $s = 0(R)$ the unexpanded result only gives additional contributions of order $\bar{R}^4\sigma^2$ so, for most practical purposes, the error in the approximate

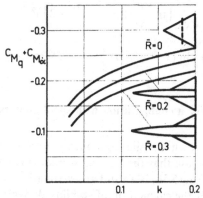

Fig. 5.3. Damping in pitch of $A = 2$ delta-wing-body combinations with $a = 0.60$.

formulae is very small. Thus the higher order effects from the fore-body can likewise be neglected, and when applying Eqs. (5.41) and (5.42) one may assume the body to taper to a point with the wing.

As an application of the theory the damping-in-pitch of $A = 2$ delta wings with bodies of various thicknesses is shown in Fig. 5.3. The bodies are cylindrical on the part covered by the wing. The wing-alone damping was calculated from Eq. (3.38) (with some higher order terms in k included). Apparently the body lowers the damping by a noticeable, but for small body radii not very large, absolute amount. For low values of k where the wing-alone damping is low (the linearized theory is then in doubt, however) the body can cause a large percentagewise reduction in damping. One should notice that the body alone was found always to give positive damping so that the reduction is an effect of the interference between wing and body.

CHAPTER 6

THE SEMI-INFINITE RECTANGULAR WING

6.1. Introduction

PROBABLY the simplest exact lifting-surface solution which retains many of the features of practical three-dimensional solutions is that for a semi-infinite rectangular wing. In supersonic flow this solution may be used to compose the exact solution for a finite rectangular wing provided the aspect ratio of the wing is so large that the Mach line from one corner does not cross the opposite side edge. In transonic flow there is always some interaction between the side edges so that the practical value of the semi-infinite-wing theory may be questioned. However, the solution is of fundamental importance for the rectangular-wing problem considered in Chapter 7 since it gives the first-order effect of finite span for a wing of arbitrary aspect ratio.

The supersonic solution was given by Stewartson (Ref. 63), Rott (Ref. 58) and Miles (Ref. 45). In principle, the transonic solution could be obtained by setting $M = 1$ in the supersonic one. In none of the papers on the super-sonic problem is the solution given in a form suitable for evaluation at $M = 1$, however, so that a separate attack on the transonic problem was instead found more practical.

6.2. Solution for the potential function

Consider a quarter-infinite rectangular wing with the leading edge located along the positive y-axis and the side edge along the positive x-axis. The boundary conditions to be satisfied at $z = 0$ are that $\varphi_z = w(x, y)$ in the first quadrant, and $\varphi = 0$ on the rest of the x,y-plane. In Refs. 45, 52 and 63 Laplace transform was used for solution of the supersonic problem but here we will use Fourier transform instead. As given in Section 1.10 the boundary value problem reads, after Fourier transformation,

$$\Phi_{yy} + \Phi_{zz} + K^2\Phi = 0 \tag{6.1}$$

$$\Phi_z(y, 0) = W(y) \quad \text{for } y > 0 \tag{6.2}$$

$$\Phi(y, 0) = 0 \quad \text{for } y < 0 \tag{6.3}$$

where
$$K = (2ku + k^2)^{1/2} \tag{6.4}$$

The proper branch of K was defined by the cut shown in Fig. 1.9.

The two-dimensional boundary value problem defined by Eqs. (6.1)–(6.4) can be solved in several different ways. In Ref. 45 Miles employed the Wiener–Hopf technique. In Refs. 63, 58 and 52 a much simpler method due to Lamb

(Ref. 28) was used which is applicable when W is independent of y. As will be shown below, however, Lamb's method can be extended to cover also cases of arbitrary W.

Introduce polar co-ordinates r, θ so that

$$y = r \cos \theta; \qquad z = r \sin \theta$$

Then a solution $\Phi = P$ satisfying Eqs. (6.1) and (6.3) and vanishing at infinity can be found by separation of variables to be

$$P = \cos (\theta/2) H_{1/2}^{(2)}(Kr) = i \cos (\theta/2) \sqrt{\left(\frac{2}{\pi Kr}\right)} e^{-iKr} \tag{6.5}$$

On the positive y-axis $P_\theta = 0$ and thus $P_z = 0$. P has a square-root singularity at the origin. From P we will construct a solution $\Phi = \Phi_0$ such that, for $y > 0$

$$\Phi_{0z}(y, 0) = e^{-isy} \tag{6.6}$$

Such a solution is given by

$$\Phi_0 = A \int_{-\infty}^{y} \exp [-is(y - y_1)] P(y_1, z) \, dy_1 \tag{6.7}$$

where A is a constant that remains to be determined.

By substitution it is easily seen that Φ_0 is a solution of Eq. (6.1). That Eq. (6.6) is also satisfied follows after calculating Φ_{0z} for $z = 0$. Since

$$\frac{\partial}{\partial z} = \frac{1}{r} \cos \theta \frac{\partial}{\partial \theta} + \sin \theta \frac{\partial}{\partial r}$$

this requires evaluation as $z \to 0$ of the integral

$$\Phi_{0z} = -A \sqrt{\left(\frac{2}{\pi K}\right)} \int_{-\infty}^{y} \exp [-is(y - y_1) - iKr_1] \left[\frac{i}{2} r_1^{-3/2} \sin (3\theta_1/2)\right.$$
$$\left. - Kr_1^{-1/2} \sin \theta_1 \cos (\theta_1/2)\right] dy_1$$

where

$$r_1 = \sqrt{(y_1{}^2 + z^2)}$$

and

$$\theta_1 = \tan^{-1} (z/y_1)$$

The contribution to the integral from the second term within the bracket apparently becomes zero in the limit. In the first term we subtract out the fundamental singularity by writing

$$\Phi_{0z}(y, 0) = -\frac{iA}{2} \sqrt{\left(\frac{2}{\pi K}\right)} e^{-isy} \lim_{z \to 0} \left\{\int_{-\infty}^{y} [\exp (isy_1 - iKr_1) - 1] r_1^{-3/2} \sin (3\theta_1/2) \, dy_1 \right.$$
$$\left. + \int_{-\infty}^{y} r_1^{-3/2} \sin (3\theta_1/2) \, dy_1\right\} = -\frac{iA}{2} \sqrt{\left(\frac{2}{\pi K}\right)} e^{-isy} (I_1 + I_2) \tag{6.8}$$

We can now directly set $z = 0$ in the first integral which then can be easily evaluated yielding

$$I_1 = -\int_{-\infty}^0 \{\exp [i(s + K)y_1] - 1\}(-y_1)^{-3/2} \, dy_1 = (1 + i)\sqrt{[2\pi(s + K)]} \quad (6.9)$$

The simplest way to calculate the second integral is to introduce the complex variables $X_1 = y_1 + iz$ and $X = y + iz$ which gives

$$I_2 = -\text{Im} \left\{\int_{-\infty + iz}^{X} X_1^{-3/2} \, dX_1\right\} = 2 \, \text{Im} \, (X^{-1/2}) \quad (6.10)$$

Hence I_2 is zero on the positive y-axis. From Eqs. (6.8)–(6.10) it then follows that Eq. (6.6) is satisfied if we choose

$$A = \tfrac{1}{2}(1 + i) \bigg/ \left(\sqrt{\frac{K}{s + K}}\right) \quad (6.11)$$

Now, it can be shown that the solution for Φ can be obtained from Φ_0 as follows:

$$\Phi = \frac{1}{2\pi} \int_{-\infty}^{\infty} \Phi_0 \, ds \int_0^{\infty} W(\eta)e^{is\eta} \, d\eta \quad (6.12)$$

For, this solution satisfies the wave equation (6.1) and is zero on the negative y-axis in accordance with Eq. (6.3). Furthermore, from Eq. (6.6) and the fundamental theorem for Fourier transforms it follows that, as required, by Eq. (6.2),

$$\Phi_z = \frac{1}{2\pi} \int_{-\infty}^{\infty} ds \int_0^{\infty} W(\eta)e^{-is(y-\eta)} \, d\eta = W(y) \quad \text{for} \quad y \geqslant 0 \quad (6.13)$$

From Eqs. (6.7), (6.11), and (6.12) we can now evaluate Φ on the positive y-axis. Thus

$$\Phi = \frac{i - 1}{2\pi} \int_{-\infty}^{\infty} \frac{ds}{\sqrt{(K + s)}} e^{-isy} \int_0^y \frac{dy_1}{\sqrt{(2\pi y_1)}} \exp [-iy_1(K - s)] \int_0^{\infty} W(\eta) \, e^{is\eta} \, d\eta \quad (6.14)$$

The order of integration may be interchanged and the integration over s carried out first. This leads to calculation of the integral

$$I = \int_{-\infty}^{\infty} \frac{\exp [-is(y - y_1 - \eta)]}{\sqrt{(K + s)}} \, ds$$

which, remembering that $\text{Im} \{K + s\} = \text{Im} \{K\} < 0$, is found to be

$$I = \begin{cases} (1 + i)\bigg/ \left(\sqrt{\dfrac{2\pi}{\eta + y_1 - y}}\right) \exp [-iK(\eta + y_1 - y)] & \text{for} \quad \eta + y_1 - y > 0 \\[6pt] 0 & \text{for} \quad \eta + y_1 - y < 0 \end{cases} \quad (6.15)$$

The final result for Φ can then be written

$$\Phi(y, +0) = -\frac{1}{\pi} \int_0^\infty W(\eta)Q(y,\eta)\,d\eta \qquad (6.16)$$

where

$$Q = \int_{y_0}^y \frac{\exp\left[-iK(2y_1 + \eta - y)\right]}{\sqrt{[y_1(y_1 - y + \eta)]}}\,dy_1 \qquad (6.17)$$

and

$$y_0 = \begin{cases} 0 & \text{for} \quad y < \eta \\ y - \eta & \text{for} \quad y > \eta \end{cases}$$

By change of integration variable to

$$t = \frac{2y_1}{|y - \eta|} - \frac{y - \eta}{|y - \eta|}$$

Q can be expressed as an "incomplete Hankel function" as follows

$$Q = \int_1^{\left|\frac{y+\eta}{y-\eta}\right|} \frac{e^{-iK|y-\eta|t}}{\sqrt{(t^2 - 1)}}\,dt \qquad (6.18)$$

For $y + \eta \to \infty$

$$Q \to -\frac{\pi i}{2} H_0^{(2)}[K|y - \eta|]$$

so that then

$$\Phi \to \frac{i}{2} \int_0^\infty W(\eta)\, H_0^{(2)}[K|y - \eta|]\,d\eta \qquad (6.19)$$

which is, of course, the result obtained if the end effect is neglected.

The inversion to the physical plane is easily carried out by aid of the following formula, valid for Im $(a) < 0$ and $x > 0$;

$$\mathscr{F}^{-1}\{\exp(-iau^{1/2})\} = \frac{1}{2}(1 + i)\frac{a}{x^{3/2}}\exp\left(-i\frac{a^2}{4x}\right) \qquad (6.20)$$

(For $x < 0$ the transform is zero.) Then, from Eqs. (6.16) and (6.18), and the Faltung theorem for Fourier integrals,

$$\varphi(x, y, +0) = -\frac{1}{\pi}\int_0^x \exp\left[-i\frac{k}{2}(x - \xi)\right]d\xi \int_0^\infty w(\xi, \eta)g(x - \xi, y, \eta)\,d\eta \qquad (6.21)$$

where

$$g(x, y, \eta) = \frac{|y - \eta|}{\sqrt{(2\pi)}}(1 + i)\sqrt{\left(\frac{k}{2}\right)}\frac{1}{x^{3/2}}\int_1^{\left|\frac{y+\eta}{y-\eta}\right|} \frac{t\,dt}{\sqrt{(t^2 - 1)}}$$

$$\times \exp\left\{-i\frac{h}{2}\left[x + \frac{t^2(y - \eta)^2}{x}\right]\right\} \qquad (6.22)$$

The integral can be evaluated by setting $t_1 = t^2 - 1$ which gives

$$g(x, y, \eta) = \frac{(1 + i)}{2x} \exp\left\{-i\frac{k}{2}\left[x + \frac{(y - \eta)^2}{x}\right]\right\}[C(\bar{v}) - iS(\bar{v})] \qquad (6.23)$$

where

$$\bar{v} = \frac{2k}{x} y\eta$$

This completes the formal solution of the problem. In the following only the case of w independent of y will be considered in detail. This case can be obtained directly from Eq. (6.21) but it is actually more convenient to calculate first the Fourier transformed potential and then perform the inversion. For this we can start from Eq. (6.16) and carry out the integration over y first, or, which is even simpler, set $s = 0$ in the expression for Φ_0, Eq. (6.6). This gives

$$\Phi(y, +0) = \frac{i - 1}{\sqrt{K}} W \int_0^y \frac{\exp(-iKy_1)}{\sqrt{(2\pi y_1)}} dy_1 = (i - 1)\frac{W}{K}[C(Ky) - iS(Ky)] \qquad (6.24)$$

The inversion may be written

$$\varphi(x, y, +0) = \int_0^x h_0(x - \xi) \exp\left[-i\frac{k}{2}(x - \xi)\right] w(\xi)\, d\xi \qquad (6.25)$$

where

$$\sqrt{(2\pi)}h_0(x) = \mathscr{F}^{-1}\left\{\frac{i - 1}{(2ku)^{1/4}}\int_0^y \frac{\exp[-i(2ku)^{1/2}y_1]}{\sqrt{2(\pi y_1)}} dy_1\right\} \qquad (6.26)$$

For computational reasons it is convenient to perform an integration by parts before evaluating the kernel. This leads to

$$\varphi(x, y, +0) = h_1(x)\exp\left(-i\frac{k}{2}x\right)w(0) - \int_0^1 h_1(x - \xi)\frac{\partial}{\partial \xi}$$

$$\times \left\{\exp\left[-i\frac{k}{2}(x - \xi)\right]w(\xi)\right\} d\xi \qquad (6.27)$$

where

$$h_1(x, y) = \int_0^x h_0(\xi, y)\, d\xi = \mathscr{F}^{-1}\left\{\frac{i + 1}{(2ku)^{1/4}}\int_0^y \frac{\exp[-i(2ku)^{1/2}y_1]}{2\pi u\sqrt{y_1}} dy_1\right\} \qquad (6.28)$$

The evaluation of h_1 will be considered in the following section.

6.3. Evaluation of h_1

Consider the inversion integral

$$h_1(x) = \frac{1}{\sqrt{(2\pi)}}\int_{-\infty}^{\infty} e^{iux} H_1(u)\, du \qquad (6.29)$$

where

$$H_1 = \frac{1 + i}{(2ku)^{1/4}} \int_0^y \frac{\exp\left[-i(2ku)^{1/2}y_1\right]}{2\pi u\sqrt{y_1}} \, dy_1 \qquad (6.30)$$

The path of integration in the complex u-plane should be taken just below the real u-axis, i.e. below the branch-point for $(2ku)^{1/2}$. To find h_1 we complete the integration path as shown in Fig. 6.1. Since no singularities are enclosed, the

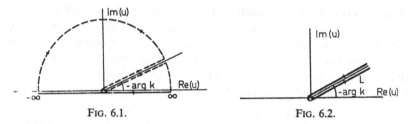

FIG. 6.1.   FIG. 6.2.

integral along this path is zero. Also, for the branch $\text{Im}\,\{(2ku)^{1/2}\} < 0$, H_1 tends to zero along the semicircle at infinity, so, by Jordan's lemma, the contribution to the integral from this part of the path is zero. Hence

$$h_1 = \frac{1}{\sqrt{2\pi}} \int_L e^{iux} H_1(u) \, du \qquad (6.31)$$

where L is shown in Fig. 6.2.

Now, for the branch $\text{Im}\,\{(2ku)^{1/2}\} < 0$, the following series representation of H_1 is convergent for all values of u, including $|u| = \infty$,

$$H_1 = \frac{2^{1/4}e^{\pi i/4}}{2\pi k^{1/4}} \sum_0^\infty \frac{(-i)^n (2k)^{n/2} y^{n+1/2} u^{n/2-5/4}}{n!\,(n + 1/2)} \qquad (6.32)$$

Hence we may substitute Eq. (6.32) in Eq. (6.31) and integrate term by term. This leads to integrals of the type

$$\int_L e^{iux} u^{n/2-5/4} \, du$$

These can be evaluated by means of the formula (see, e.g. Ref. 13, p. 14)

$$\int_L u^{-\alpha} e^{-u\delta} \, ds = \frac{2\pi i(-\delta)^{\alpha-1}}{\Gamma(\alpha)} \qquad (6.33)$$

$$-(\pi/2 - \arg k) < \arg \delta < \pi/2 + \arg k$$

Then, setting $\delta = -ix$, and with $\arg k < 0$,

$$h_1 = \frac{2^{1/4}e^{\pi i/4}}{k^{1/4}\sqrt{2\pi}} \sum_0^\infty \frac{e^{5\pi i/8 - \pi in/4} x^{1/4-n/2} (2k)^{n/2} y^{1/2+n}}{n!\,(n + 1/2)\Gamma(5/4 - n/2)} \qquad (6.34)$$

With $v = y\sqrt{(k/x)}$ the result may be written as follows, after some simplification,

$$h_1 = \sqrt{\left(\frac{x}{\pi k}\right)} g_1(v) \qquad (6.35)$$

where

$$g_1 = g_1' + ig_1'' = -\frac{(2i)^{-1/4}}{\Gamma(1/4)}(a_1 + ib_1) + \frac{(2i)^{1/4}}{\Gamma(3/4)}(c_1 + id_1) \qquad (6.36)$$

and

$$a_1 = v^{1/2}\left[8 - \sum_1^\infty \frac{(-1)^r 3 \cdot 7 \cdot 11 \ldots (8r - 5)}{(4r)! \, (4r + 1/2)4^{r-1}} v^{4r}\right] \qquad (6.37)$$

$$b_1 = v^{5/2}\left[\frac{2}{5} + \sum_1^\infty \frac{2(-1)^r 3 \cdot 7 \cdot 11 \ldots (8r - 1)}{(4r + 2)! \, (4r + 5/2)4^r} v^{4r}\right] \qquad (6.38)$$

$$c_1 = v^{3/2}\left[\frac{2}{3} + \sum_1^\infty \frac{(-1)^r 1 \cdot 5 \cdot 9 \cdot 13 \ldots (8r - 3)}{(4r + 1)! \, (4r + 3/2)4^r} v^{4r}\right] \qquad (6.39)$$

$$d_1 = -v^{7/2} \sum_1^\infty \frac{2(-1)^r 1 \cdot 5 \cdot 9 \cdot 13 \ldots (8r + 1)}{(4r + 3)! \, (4r + 7/2)4^{r+1}} v^{4r} \qquad (6.40)$$

Using this series expansion the functions g_1' and g_1'' were tabulated for real values of $\sqrt{v}$ by aid of the Swedish electronic high-speed computer BESK. The results are shown in Table 1 and Fig. 6.4. The series were broken off when

TABLE 1

$\sqrt{v}$	g_1'	g_1''	$\sqrt{v}$	g_1'	g_1''
0.0	0.00000	0.00000	1.3	−1.07776	1.04822
0.1	−0.17083	0.07125	1.4	−1.02685	1.06269
0.2	−0.33808	0.14396	1.5	−0.98439	1.04816
0.3	−0.49821	0.21949	1.6	−0.96566	1.01514
0.4	−0.64779	0.29895	1.7	−0.97652	0.98499
0.5	−0.78345	0.38313	1.8	−1.00297	0.97945
0.6	−0.90201	0.47227	1.9	−1.01547	0.99846
0.7	−1.00045	0.56588	2.0	−1.00155	1.01159
0.8	−1.07606	0.66254	2.1	−0.99113	0.99941
0.9	−1.12659	0.75959	2.2	−1.00338	0.99406
1.0	−1.15062	0.85294	2.3	−1.00163	1.00499
1.1	−1.14822	0.93690	2.4	−0.99697	0.99721
1.2	−1.12186	1.00446	2.5	−1.00274	1.00168

the magnitude of individual terms became smaller than 2^{-20} ($\sim 10^{-6}$). For $\sqrt{v} > 2.6$ some term exceeded in magnitude 2^{20} which was outside the digital capacity of the BESK, and the computation was stopped.

6.4. Asymptotic formula for g_1 for large values of v

For large values of $(ku)^{1/2}y$ an asymptotic expansion of H_1 gives

$$H_1 \sim \frac{1}{2u^{3/2}\sqrt{(\pi k)}} + \frac{(i-1)}{2\pi\sqrt{(y)(2k)^{3/4}u^{7/4}}} \exp\left[-i(2ku)^{1/2}y\right] \qquad (6.41)$$

The first term is easily inverted to yield (see Chapter 2)

$$-\left(\frac{x}{2\pi ik}\right)^{1/2} = \sqrt{\left(\frac{x}{\pi k}\right)}(i-1) \qquad (6.42)$$

so that the limiting value of g_1 as $v \to \infty$ is

$$g_1 \to i - 1 \qquad (6.43)$$

Table 1 shows that this value is reached within 0.5 per cent in magnitude already for $\sqrt{v} \geqslant 2.3$. For the second term consider the inversion integral

$$\frac{1}{\sqrt{(2\pi)}} \int_{-\infty}^{\infty} q(u) \exp\left[iux - i(2ku)^{1/2}y\right] du \qquad (6.44)$$

where

$$q(u) = \frac{i-1}{2\pi\sqrt{(y)(2k)^{3/4}u^{7/4}}}$$

Let k be purely real. Then the cut in the u-plane will be the positive real u-axis and the integration is to be taken on the lower side of the cut. In order to avoid the singularity at $u = 0$, the path of integration is deformed as shown in Fig. 6.3.

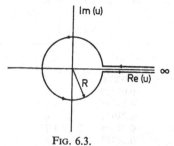

FIG. 6.3.

For reasons to be apparent later we chose $R = \beta k y^2/2x^2$, where $0 < \beta < 1$, say $\beta = 1/2$. Consider first the part of the integral along the lower Re (u)-axis

$$\frac{1}{\sqrt{(2\pi)}} \int_{R}^{\infty} q(u)e^{i\alpha p(u)} du \qquad (6.45)$$

where

$$p(u) = \frac{x}{y\sqrt{k}}u - \sqrt{2u}$$

and

$$\alpha = y\sqrt{k}$$

As α becomes large this integral can be estimated by the method of stationary phase (see Ref. 14, p. 51). The phase is stationary, i.e. $p' = 0$, for $u = u_0 = ky^2/2x^2$. The main contribution to the integral comes from the region around $u = u_0$. Hence the above choice of R so that $R < u_0$.

Now, as $\alpha \to \infty$ (Ref. 14):

$$\frac{1}{\sqrt{(2\pi)}} \int_R^\infty q(u)e^{i\alpha p(u)}\, du \sim \frac{1}{[\alpha p''(u_0)]^{1/2}} q(u_0) \exp\left[i\alpha p(u_0) + \frac{\pi i}{4}\right] \quad (6.46)$$

Substituting for the functions p and q then gives, after simplification,

$$\frac{1}{\sqrt{2\pi}} \int_R^\infty q(u)e^{i\alpha p(u)}\, du \sim \sqrt{\left(\frac{x}{\pi k}\right)} \sqrt{\left(\frac{2}{\pi}\right)} v^{-3} \exp\left(-i\frac{v^2}{2}\right) \quad (6.47)$$

Turning now to the remainder of the integral we see that, since, for the proper branch, $\arg u < 0$, the integrand goes exponentially to zero as $\sqrt{(ky)} \to \infty$,

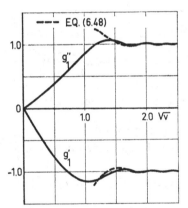

FIG. 6.4. The function $g_1 = g_1' + ig_1''$.

even in the lower half-plane with the particular choice of R. The only contribution of the order v^{-n} may then come from the part of the circle where $\arg u \to -0$. A calculation shows that this contribution will be of the order $\sqrt{(x/\pi k)}v^{-4}$, i.e. of higher order than that of Eq. (6.47). Hence we have found the following asymptotic expression as $v \to \infty$

$$g_1 \sim i - 1 + \sqrt{\left(\frac{2}{\pi}\right)} v^{-3} \exp\left(-i\frac{v^2}{2}\right) \quad (6.48)$$

From Eq. (6.48) it is evident that g_1 is convergent also for $v^2 = +\infty$; Im $(v^2) = 0$, so that k from now on can be considered purely real.

Results according to this formula are shown in Fig. 6.4 together with the exact values.

6.5. Calculation of sectional forces and moments

The sectional forces and moments for translational and pitching oscillations, $l_{11} - l_{22}$, which are defined in Section 1.7 are readily obtained by introducing

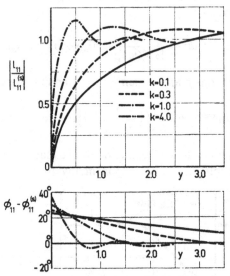

FIG. 6.5. Sectional lift due to translation (divided by strip-theory value) vs. y.

the result for φ into Eqs. (1.60)–(1.64). By means of a number of integration by parts the calculation can be reduced to the evaluation of the following integrals

$$l_{ij} = 4(\pi k)^{-1/2}\left[m_{ij}h_1(1, y) \exp\left(-i\frac{k}{2}\right) - \int_0^1 h_1(x, y) \exp\left(-i\frac{k}{2}x\right) M_{ij}(x)\, dx \right]$$

where   (6.49)

$$m_{11} = m_{12} = ik$$

$$m_{21} = m_{22} = 1$$

$$M_{11} = \frac{3}{2} k^2 + \frac{ik^3}{2} (1 - x)$$

$$M_{12} = ik + \frac{3}{2} k^2 x + \frac{ik^3}{4} (1 - x^2)$$

(6.50)

$$M_{21} = -\frac{5}{2} ik + 2k^2(1 - x) + \frac{ik^3}{4} (1 - x)^2$$

$$M_{22} = 1 - \frac{5}{2} ikx + k^2(1 - x^2) + \frac{ik^3}{12} (2 - 3x + x^3)$$

The two-dimensional (strip-theory) values of l_{ij}, $l_{ij}^{(s)}$, are obtained by setting $h_1 = \sqrt{(x/\pi k)}(i-1)$ in Eq. (6.49). Instead of calculating l_{ij} directly it is sometimes convenient to calculate instead the end corrections due to three-dimensional flow defined by Eq. (1.65). For $y = 0$, $l_{ij} = 0$ so that $\Delta l_{ij} = -l_{ij}^{(s)}$ and hence $c_{ij} = -1$. For large y, Δl_{ij} and c_{ij} tend to zero.

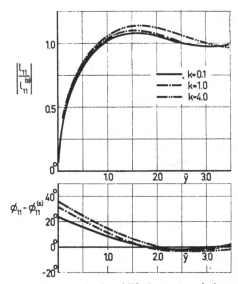

FIG. 6.6. Sectional lift due to translation vs. $\bar{y}$.

In Ref. 36 were given tables of $\Delta l_{ij}(y)$ for some values of k, calculated on the BESK. In Figs. 6.5 and 6.6 are shown some results for $l_{11}/l_{11}^{(s)} = 1 + c_{11}$. When plotted against the reduced variable $\bar{y} = y\sqrt{(k + k^2/2)}$ as in Fig. 6.6 (the choice of this variable was made on a semi-empirical basis) the results for different k are seen to fall on approximately the same curve. The other coefficients show the same tendency. From Fig. 6.6 it then follows that strip-theory gives an error of less than 5 per cent in vector magnitude and 5° in phase angle for $\bar{y}$ larger than about 2.5.

6.6. Approximate results for rectangular wings of large aspect ratios

From the results of the previous section it is evident that, for a rectangular wing with large aspect ratio (or reduced frequency), the flow could be calculated with good approximation by neglecting the mutual flow interference between the edges and hence assume the edges to act independently. With the side edges at $y = \pm \sigma$ we then get immediately, by use of the results of the previous section, the following approximate sectional forces on the finite wing:

$$l_{ij} \approx l_{ij}^{(s)} + \Delta l_{ij}(\sigma - y) + \Delta l_{ij}(\sigma + y) \qquad (6.51)$$

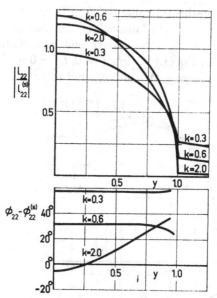

FIG. 6.7. Approximate sectional moment due to pitch for an $A = 2$ rectangular wing at $M = 1$.

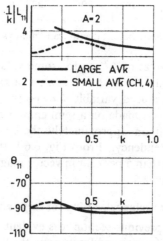

FIG. 6.8. Total lift due to translation for an $A = 2$ rectangular wing at $M = 1$.

In Fig. 6.7 are shown results obtained from Eq. (6.51) for l_{22} (sectional moment due to pitch) on an $A = 2$ wing oscillating with $k = 0.3$, $k = 0.6$ and $k = 2.0$. The error of the approximate theory is indicated by the amount of moment that persists for $|y| \geqslant \sigma$. Apparently it is not very satisfactory for the lowest value of k but is fairly good for $k = 0.6$ and good for $k = 2.0$. In Fig. 6.8 the result for L_{11} (total lift due to translation) for a wing of $A = 2$ is compared to that obtained from the low-aspect-ratio theory of Chapter 4. The very good agreement in the frequency region around $k = 0.5$ is surprising in view of the quite different assumptions underlying the two theories. However, as shown in Ref. 36, the agreement for moment coefficients is not as good.

The more precise limitations of the present approximate theory are discussed in Chapter 7.

CHAPTER 7

THE RECTANGULAR WING OF ARBITRARY
ASPECT RATIO WITH CONTROL SURFACE

7.1. Introduction

IN THIS chapter we will describe the exact solution for the oscillating rectangular wing of arbitrary aspect ratio as given in Ref. 38. The solution is based on Schwarzschild's method (Ref. 60), which was originally developed for the problem of electromagnetic diffraction by a two-dimensional finite slit. This method was applied by Gunn (Ref. 20) to the problem of a low aspect ratio rectangular wing at steady angle of attack in a supersonic flow, but for the corresponding unsteady-flow problem it has generally been considered too complicated to be practical. For transonic flow the method is quite feasible, however, particularly for adaptation to automatic computation.

7.2. Principle of solution

Let the finite rectangular wing occupy the region $0 \leqslant x \leqslant 1$, $|y| \leqslant \sigma$ on $z = 0$. The Fourier-transformed boundary value problem for the wing as given in Section 1.8 reads (see Fig. 7.1)

$$\Phi_{yy} + \Phi_{zz} + K^2\Phi = 0 \tag{7.1}$$

$$\Phi_z(y, 0) = W \quad \text{for} \quad |y| < \sigma \tag{7.2}$$

$$\Phi(y, 0) = 0 \quad \text{for} \quad |y| \geqslant \sigma \tag{7.3}$$

where

$$K = (2ku + k^2)^{1/2}$$

This is the classical boundary value problem of two-dimensional diffraction around a finite strip. Following Schwarzschild's (Ref. 60) method of solution

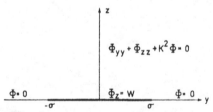

FIG. 7.1. Boundary value problem after Fourier transformation.

we first obtain a first-order solution $\Phi^{(0)}$ by assuming the wing span to be infinite. Then it is necessary to assume some distribution of W outside the strip. This distribution is actually arbitrary as long as it gives a finite $\Phi^{(0)}$. If W is independent of y on the wing it is convenient to let it remain constant on the whole y-axis. This then gives for $\Phi^{(0)}$ the solution

$$\Phi^{(0)} = \frac{iW}{K} e^{-iKz} \tag{7.4}$$

i.e. the strip theory solution considered in Chapter 2.

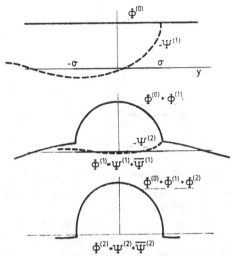

FIG. 7.2. Principle of solution.

One can also set $W = 0$ outside the wing strip which gives for the Kirchhoff approximation. Whatever distribution is assumed outside the wing strip, the solution is fairly easily obtained from Green's formula, which gives

$$\Phi^{(0)} = \frac{i}{2} \int_{-\infty}^{\infty} H_0^{(2)}(K\tilde{r}) W(\eta) \, d\eta \tag{7.5}$$

where

$$\tilde{r} = \sqrt{[(y - \eta)^2 + z^2]}$$

We now have a solution of Eq. (7.1) that fulfills all the boundary conditions except that given by Eq. (7.3). Hence it remains to add a solution that cancels $\Phi^{(0)}$ for $|y| \geqslant \sigma$ but has a zero normal derivative for $|y| < \sigma$. This is obtained by means of a step-by-step procedure by considering one edge at a time as shown schematically in Fig. 7.2. First a solution $\Psi^{(1)}$ is found which cancels $\Phi^{(0)}$ to the right of the starboard side edge but leaves the normal derivative to the left of the edge unchanged. A corresponding solution $\overline{\Psi}^{(1)}$ for the port edge is also added. This gives an improved solution $\Phi = \Phi^{(0)} + \Phi^{(1)}$, where $\Phi^{(1)} = \Psi^{(1)} + \overline{\Psi}^{(1)}$. This still does not quite vanish outside the wing strip.

Therefore a new solution $\Psi'^{(2)}$ is calculated which cancels $\Phi^{(1)}$ to the right of
the side edge but has a zero normal derivative to the left of the edge, etc. The
process is continued in this way until the desired accuracy is obtained.
Schwarzschild (Ref. 60) showed that for the case of W independent of y the
series obtained in this manner converges for all non-zero values of KA when K

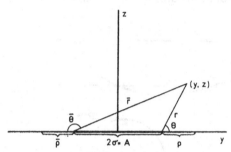

FIG. 7.3. Auxiliary co-ordinate system.

is a real number. In Ref. 38 the proof was extended to show that the series for
Φ, as well as for φ, is convergent for any finite $w(x, y)$ distribution on the wing,
provided A and k are both different from zero.

There now remains the auxiliary problem to calculate the potential functions
$\Psi'^{(n)}$ and $\bar{\Psi}'^{(n)}$. For this it is convenient to introduce two sets of polar co-
ordinates r, θ and $\bar{r}$, $\bar{\theta}$ with origins at the starboard and port edges, respectively
(see Fig. 7.3). On the wing strip $\theta = \bar{\theta} = \pi$. The distances from a point on
the y-axis outside the wing strip to the nearest edge are denoted by ϱ and $\bar{\varrho}$,
respectively (these are used as integration variables). The solutions for $\Psi'^{(n)}$ and
$\bar{\Psi}'^{(n)}$ may then be formally written as follows:

$$\Psi'^{(n)}(r, \theta) = -\frac{1}{\pi} \int_0^\infty G(r, \varrho, \theta)\Phi^{(n-1)}(\varrho + \sigma, + 0) \, d\varrho \qquad (7.6)$$

$$\bar{\Psi}'^{(n)}(\bar{r}, \bar{\theta}) = -\frac{1}{\pi} \int_0^\infty G(\bar{r}, \bar{\varrho}, \bar{\theta})\Phi^{(n-1)}(-\bar{\varrho} - \sigma, +0) \, d\bar{\varrho} \qquad (7.7)$$

Or, for $n \geqslant 2$,

$$\Psi'^{(n)}(r, \theta) = -\frac{1}{\pi} \int_0^\infty G(r, \varrho, 0)\bar{\Psi}'^{(n-1)}(\varrho + A, \pi) \, d\varrho \qquad (7.8)$$

$$\bar{\Psi}'^{(n)}(\bar{r}, \bar{\theta}) = -\frac{1}{\pi} \int_0^\infty G(\bar{r}, \bar{\varrho}, \bar{\theta}) \Psi'^{(n-1)}(\bar{\varrho} + A, \pi) \, d\varrho \qquad (7.9)$$

The calculation of the Green's function $G(r, \varrho, \theta)$ will be the subject of the
following section.

7.3. Solution of the auxiliary problem

In Ref. 60 G was found by means of Green's formula. Here a somewhat simpler derivation will be presented. The problem to be solved for the starboard side edge reads as follows:

$$\Psi_{rr} + \frac{1}{r}\Psi_r + \frac{1}{r^2}\Psi_{\theta\theta} + K^2\Psi = 0 \tag{7.10}$$

$$\Psi_\theta(r, \pi) = 0 \tag{7.11}$$

$$\Psi(r, 0) = F(r) \tag{7.12}$$

An elementary solution $\Psi = Q$ of Eq. (7.10) that vanishes at infinity can be found by separation of variables to be

$$Q = \sin(\theta/2)H^{(2)}_{1/2}(Kr) = -i\sin(\theta/2)\sqrt{\left(\frac{2}{\pi Kr}\right)}e^{-iKr} \tag{7.13}$$

This solution fulfills Eq. (7.11) but is zero for $\theta = 0$. Returning for a moment to the original co-ordinate system y, z a solution $\Psi = \Psi_0$ is therefore constructed from $Q(y, z)$ as follows

$$\Psi_0 = C\int_{-\infty}^{y} e^{-ip(y-y_1)}Q(y_1, z)\,dy_1 \tag{7.14}$$

where p is a real number. By substitution it is easily seen that Ψ_0 is a solution of Eq. (7.1). Also $\Psi_{0_z}(y, 0) = 0$ to the left of the edge since $Q_0(r, \pi) = 0$. By introducing Eq. (7.14) into Eq. (7.13) we find that, for $\theta = 0$,

$$\Psi_0(r, 0) = -iCe^{-ipr}\int_0^\infty \sqrt{\left(\frac{2}{\pi Kr_1}\right)}e^{-ir_1(K+p)}\,dr_1 \tag{7.15}$$

since $Q(r, 0) = 0$. Hence $\Psi_0(r, 0)$ is proportional to e^{-ipr}. We will determine C so that the proportionality constant becomes equal to unity. Carrying out the integration in Eq. (7.15) we then find that

$$C = \frac{i-1}{2}\sqrt{[K(K+p)]} \tag{7.16}$$

Hence

$$\Psi_0(r, \pi) = (1+i)\sqrt{(K+p)}e^{ipr}\int_r^\infty \frac{e^{-ir_1(K+p)}}{\sqrt{(2\pi r_1)}}\,dr_1$$

or

$$\Psi_0(r, \pi) = -(1+i)e^{ipr}\left[C(K_1r) - iS(K_1r) - \frac{1}{2} + \frac{i}{2}\right] \tag{7.17}$$

where

$$K_1 = K + p \tag{7.18}$$

The solution for Ψ is obtained from Ψ_0 by superposition, setting

$$\Psi = \frac{1}{\sqrt{(2\pi)}} \int_{-\infty}^{\infty} H(p)\,\Psi_0(p)\,dp \qquad (7.19)$$

$H(p)$ is to be determined so that Eq. (7.12) is fulfilled. This leads to the following integral equation

$$\frac{1}{\sqrt{(2\pi)}} \int_{-\infty}^{\infty} \Psi_0(r,0)H(p)\,dp \equiv \frac{1}{\sqrt{(2\pi)}} \int_{-\infty}^{\infty} e^{-ipr}H(p)\,dp = F(r) \qquad (7.20)$$

Now assume for a moment that $F(r)$ is given also for negative values of r. Then Eq. (7.20) may be solved by means of Fourier transforms. Applying the inversion formula gives directly

$$H(p) = \frac{1}{\sqrt{(2\pi)}} \int_{-\infty}^{\infty} F(\varrho)\,e^{ip\varrho}\,d\varrho \qquad (7.21)$$

Hence, by introducing this into Eq. (7.19),

$$\Psi(r,\pi) = \frac{1}{2\pi} \int_{-\infty}^{\infty} \int_{-\infty}^{\infty} \Psi_0(r,\pi,p)F(\varrho)\,e^{ip\varrho}\,d\varrho\,dp \qquad (7.22)$$

In this we may interchange the order of integration and carry out that over p first. Introducing Eq. (7.17), changing integration variable to K_1, and integrating by parts gives

$$\int_{-\infty}^{\infty} \Psi_0 e^{ip\varrho}\,dp = (i-1)\frac{\sqrt{r}}{r+\varrho}\,e^{-iK(r+\varrho)} \int_{-\infty}^{\infty} \frac{e^{iK_1\varrho}}{(2\pi K_1)^{1/2}}\,dK_1 \qquad (7.23)$$

Keeping in mind the original assumption that $\mathrm{Im}\,(K) = \mathrm{Im}\,(K_1) < 0$ the integral is found to be:

$$\Psi_0 e^{ip\varrho}\,d\varrho = \begin{cases} 2\sqrt{\left(\dfrac{r}{\varrho}\right)}\dfrac{e^{-iK(r+\varrho)}}{r+\varrho} & \text{for} \quad \varrho > 0 \\[2mm] 0 & \text{for} \quad \varrho < 0 \end{cases} \qquad (7.24)$$

Introducing this into Eq. (7.22) gives, finally,

$$\Psi(r,\pi) = \frac{1}{\pi} \int_{0}^{\infty} \sqrt{\left(\frac{r}{\varrho}\right)}\frac{e^{-iK(r+\varrho)}}{r+\varrho}F(\varrho)\,d\varrho \qquad (7.25)$$

and, consequently,

$$G(r,\varrho,\pi) = \sqrt{\left(\frac{r}{\varrho}\right)}\frac{e^{-iK(r+\varrho)}}{r+\varrho} \qquad (7.26)$$

It is possible to find G for other values of θ accordingly but this seems to be of less interest for the present problem.

7.4. Complete solution for the general case

The complete solution may now be given in the physical co-ordinates by aid of the following two inversion formulas valid for $x > 0$ (for $x < 0$ the inverse transforms are zero), $\text{Im}\,(k) < 0$ and a real

$$\mathscr{F}^{-1}\{H_0^{(2)}(Ka)\} = \sqrt{\left(\frac{2}{\pi}\right)}\frac{i}{x}\exp\left[-i\frac{k}{2}\left(x + \frac{a^2}{x}\right)\right] \tag{7.27}$$

$$\mathscr{F}^{-1}\{e^{-iKa}\} = (1+i)\sqrt{\left(\frac{k}{2}\right)}\frac{a}{x^{3/2}}\exp\left[-i\frac{k}{2}\left(x + \frac{a^2}{x}\right)\right] \tag{7.28}$$

Thus the final result for the potential distribution on the wing reads

$$\varphi = \sum_0^\infty \varphi^{(n)} \tag{7.29}$$

$$\varphi^{(n)} = \psi^{(n)} + \bar{\psi}^{(n)} \quad \text{for} \quad n > 0 \tag{7.30}$$

$$\varphi^{(0)}(x, y) = -\frac{1}{2\pi}\int_{-\infty}^{\infty} d\eta \int_0^x \frac{d\xi}{x-\xi}\exp\left\{-i\frac{k}{2}\left[x - \xi + \frac{(y-\eta)^2}{x-\xi}\right]\right\}w(\xi, \eta) \tag{7.31}$$

$$\psi^{(n)}(x, r) = \frac{(i-1)\sqrt{(k)}}{2\pi^{3/2}}\int_0^\infty \sqrt{\left(\frac{r}{\varrho}\right)}\,d\varrho \int_0^x \frac{d\xi}{(x-\xi)^{3/2}}$$

$$\times \exp\left\{-i\frac{k}{2}\left[x - \xi + \frac{(r+\varrho)^2}{x-\xi}\right]\right\}\bar{\psi}^{(n-1)}(\xi, A + \varrho) \tag{7.32}$$

$$\bar{\psi}^{(n)}(x, \bar{r}) = \frac{(i-1)\sqrt{(k)}}{2\pi^{3/2}}\int_0^\infty \sqrt{\left(\frac{\bar{r}}{\bar{\varrho}}\right)}\,d\bar{\varrho} \int_0^x \frac{d\xi}{(x-\xi)^{3/2}}$$

$$\times \exp\left\{-i\frac{k}{2}\left[x - \xi + \frac{(\bar{r}+\bar{\varrho})^2}{x-\xi}\right]\right\}\psi^{(n-1)}(\xi, A + \bar{\varrho}) \tag{7.33}$$

For $n = 1$ $\psi^{(0)}$ and $\bar{\psi}^{(0)}$ in Eqs. (7.32) and (7.33) are to be replaced by $\varphi^{(0)}$.

7.5. Calculation for rigid translational and pitching oscillations

Although the formulas Eqs. (7.29)–(7.33), given in the physical plane, may be directly applied it is actually more convenient to carry out the calculation in the transformed plane before inversion.

Since W for the modes considered is independent of y we choose accordingly as solution for $\Phi^{(0)}$ that given by Eq. (7.4). Thus, for $z = +0$,

$$\Phi^{(0)} = \frac{iW}{K} \tag{7.34}$$

The inversion of this is straightforward and gives as a result the strip-theory solution (Ref. 50)

$$\varphi^{(0)} = \varphi^{(s)} = \frac{i-1}{2} \int_0^x \frac{\exp\left[-i\frac{k}{2}(x-\xi)\right]}{\sqrt{[\pi k(x-\xi)]}} w(\xi)\, d\xi \qquad (7.35)$$

The calculation of $\Psi'^{(1)}$ requires the following integral formula which may be proved by differentiation with respect to K.

$$\frac{1}{\pi}\int_0^\infty \sqrt{\left(\frac{r}{\varrho}\right)} \frac{e^{-iK(r+\varrho)}}{r+\varrho}\, d\varrho = -(1+i)\left[C(Kr) - iS(Kr) - \frac{1}{2} + \frac{i}{2}\right] \qquad (7.36)$$

Thus

$$\Psi'^{(1)}(r) = \bar{\Psi}'^{(1)}(r) = \frac{i-1}{K}\left[C(Kr) - iS(Kr) - \frac{1}{2} + \frac{i}{2}\right] W \qquad (7.37)$$

$\Phi^{(0)} + \Psi'^{(1)}$ gives, as it indeed should, exactly the same result as that for a semi-infinite rectangular wing given in Chapter 6. Let $k_{ij}^{(n)}$ denote the sectional coefficients l_{ij} corresponding to $\Psi'^{(n)}$. By use of the results of Chapter 6 the coefficients $k_{ij}^{(1)}$ are then immediately found to be:

$$k_{ij}^{(1)} = 4(\pi k)^{-1/2}\left\{m_{ij}k_1(1,r) \exp\left(-i\frac{k}{2}\right) - \int_0^1 k_1(x,r) \exp\left(-i\frac{k}{2}x\right) M_{ij}(x)\, dx\right\} \qquad (7.38)$$

where m_{ij} and M_{ij} are defined by Eq. (6.50) and where

$$k_1 = \sqrt{\left(\frac{x}{\pi k}\right)}[g_1(v) + 1 - i] = h_1(x) - h_1(\infty) \qquad (7.39)$$

$$v = r\sqrt{\left(\frac{k}{x}\right)}$$

with $g_1(v)$ given by the series

$$g_1 = \sum_0^\infty \frac{\exp\left[(\pi i/4)(7/2 - n)\right] 2^{n/2-1/4}v^{1/2+n}}{n!\,(n+1/2)\Gamma(5/4 - n/2)} \qquad (7.40)$$

The approximate theory described in Chapter 6 for wings of large aspect ratio is actually identical to the present one if the series (7.29) is broken off after the second term. Since this approximate theory turned out to give good results even for fairly low values of $A\sqrt{k}$ it is suggested that the third and higher order terms are small in many practical cases. Therefore only the asymptotic value of $\Psi'^{(2)}$ for large values of $A\sqrt{k}$ is calculated. As seen below this considerably simplifies the analysis. From Eq. (7.37) we find for large values of Kr

$$\bar{\Psi}'^{(1)}(r) = \frac{1-i}{\sqrt{K}} \int_r^\infty \frac{e^{-iKu}\, du}{\sqrt{(2\pi u)}} \sim -\frac{1+i}{K^{3/2}} \frac{e^{-iKr}}{\sqrt{(2\pi r)}} \qquad (7.41)$$

When introducing this into Eqs. (7.8) and (7.26) for the determination of $\Psi'^{(2)}$ the terms in the integral containing u may be written

$$\overline{\Psi}'^{(1)}(\varrho + A)e^{-iK(r+\varrho)} \approx \sqrt{\left(\frac{2\varrho + A + r}{\varrho + A}\right)}\overline{\Psi}'^{(1)}(2\varrho + r + A) \qquad (7.42)$$

Hence the following approximate result is obtained

$$\Psi'^{(2)}(r) \approx -\frac{1}{\pi}\int_0^\infty \sqrt{\left[\frac{r(2\varrho + A + r)}{\varrho(\varrho + A)}\right]}\frac{\Psi'^{(1)}(2\varrho + A + r)}{r + \varrho}\,d\varrho \qquad (7.43)$$

Since u only enters in $\Psi'^{(1)}$ and $\Psi'^{(2)}$ the inversion of Eq. (7.43) is obtained simply by replacing these by their inverse transforms. Also the associated aerodynamic coefficients may be directly given as follows†

$$k_{ij}^{(2)} \approx -\frac{1}{\pi}\int_0^\infty \sqrt{\left[\frac{r(2\varrho + A + r)}{\varrho(\varrho + A)}\right]}\frac{k_{ij}^{(1)}(2\varrho + A + r)}{r + \varrho}\,d\varrho \qquad (7.44)$$

This integral has to be evaluated numerically. Some tables of $c_{ij}^{(2)}$ were given in Ref. 38.

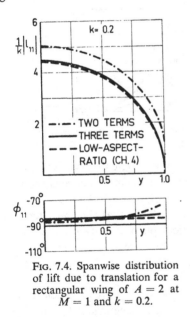

FIG. 7.4. Spanwise distribution of lift due to translation for a rectangular wing of $A = 2$ at $M = 1$ and $k = 0.2$.

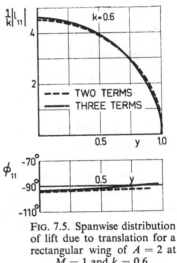

FIG. 7.5. Spanwise distribution of lift due to translation for a rectangular wing of $A = 2$ at $M = 1$ and $k = 0.6$.

In order to test the convergence of the method some results for an $A = 2$ wing are computed. This is an aspect ratio typical for wings of practical interest and is low enough to give a critical test of the convergence. In Figs. 7.4 and 7.5 results for l_{11}, i.e. distribution of lift due to translation, are shown for $k = 0.2$

† This approximation is also valid for the supersonic case.

and $k = 0.6$, respectively. For $k = 0.2$ the parameter $A\sqrt{k}$ is only 0.9 so that one would expect the convergence to be poor in this case. Nevertheless the third term gives a fairly small contribution compared to the second term. This indicates that the fourth term probably is very small. However, since the third term was calculated only approximately assuming $A\sqrt{k}$ to be large, the error in this may be larger than the neglected fourth term (an order of magnitude analysis shows that they actually should be of the same order). In Fig. 7.4 is therefore

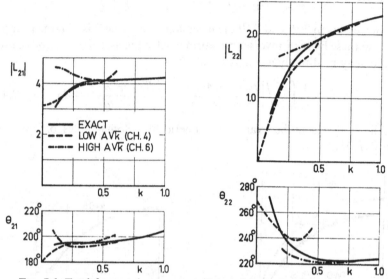

FIG. 7.6. Total force and moment coefficient for an $A = 2$ rectangular wing oscillating in pitch at $M = 1$.

included as an additional check results from the approximate theory of Chapter 4 which is valid for low values of $A\sqrt{k}$. As seen the agreement between the two theories is excellent which serves as a valuable check on both theories.

For the higher frequency of $k = 0.6$ Fig. 7.5 shows that the third term is small and probably negligible for most practical purposes. Hence the high-aspect-ratio theory of Chapter 6 is applicable in this case.

We are now in a position to determine more generally the ranges of applicability of the approximate theories described in Chapters 4 and 6 as well as the present three-term solution. For this purpose total-force-coefficients for pitching oscillations are plotted in Fig. 7.6. For the lift-coefficient L_{21}, all three theories agree well in a small region near $k = 0.4$. The low-aspect-ratio theory of Chapter 4 agrees very well with the present one for values of k between 0.15 and 0.50. For $k < 0.15$ more than three terms are required.

For the moment coefficient L_{22}, the low aspect ratio theory begins to deviate from the present solution at a much lower k. This is due to the spurious pressure singularities at the leading edge introduced by the low aspect ratio theory. None the less there is a fair amount of overlapping with the present theory in the region around $k = 0.2$. Allowing an error of 10 per cent in vector magnitude and $5°$ in phase angle we may therefore give the limits of applicability of the three theories as follows:

Present three-term solution: $A\sqrt{k} > 0.8$.
High aspect ratio theory of Chapter 6 (present two-term solution): $A\sqrt{k} > 1.3$.
Low aspect ratio theory of Chapter 4: $A\sqrt{k} < 1.0$.

The limits are set primarily by the moment coefficients while those for the lift coefficients are less restrictive. For translational oscillations the limits are about the same, but for flexible modes they might be entirely different. Thus the low aspect ratio solution is probably more restricted for modes involving chordwise deformations since then chordwise flow derivatives are larger, making higher order terms in the theory of greater importance. Conversely, for modes with spanwise deformation the low aspect ratio theory would be applicable for higher values of $A\sqrt{k}$. The present three-term solution, on the other hand, would have the opposite properties. Hence these two theories complement each other so that any combination of k and A may be treated by one of them. Now, as discussed in Chapter 1, the linearized theory ceases to be valid for very low values of k so that the three-term solution probably covers the whole range of reduced frequencies for which linearization is admissible, except for wings of extremely low aspect ratios $(A < 1)$.

The above estimates were made for low values of k. For higher values of k they can also be used if $\sqrt{k}$ is replaced by $\sqrt{(k + k^2/2)}$.

7.6. Approximate expressions for total forces and moments for large values of $A\sqrt{k}$

When calculating the total force and moment coefficients for such large values of $A\sqrt{k}$ that the two-term solution can be used, the following integral needs to be evaluated:

$$I = \int_0^A \Psi^{(1)}(r) \, dr = W \frac{(i-1)}{K} \int_0^A \left[C(Kr) - iS(Kr) - \frac{1}{2} + \frac{i}{2} \right] dr \quad (7.45)$$

It can be written

$$I = \int_0^\infty \Psi^{(1)}(r) \, dr - \int_A^\infty \Psi^{(1)}(r) \, dr \quad (7.46)$$

For large values of KA the last term is of the order $WK^{-5/2}A^{-1/2}$, which turns out to be the same as that of the contribution from $\Psi^{(2)}$. Hence it could be similarly neglected so that

$$I \approx \int_0^\infty \Psi^{(1)}(r) \, dr = -\frac{W}{2K^2} \quad (7.47)$$

Now the inversion of this result is simply

$$\int_0^\infty \psi^{(1)}(x, r)\, dr = -\frac{i}{4k} \int_0^x \exp\left[-i\frac{k}{2}(x - \xi)\right] w(\xi)\, d\xi \qquad (7.48)$$

The results for the coefficients may then be written as follows, after some elementary calculations:

$$L_{ij} = L_{ij}^{(s)} + \Delta L_{ij} \qquad (7.49)$$

where $L_{ij}^{(s)}$ is the corresponding strip theory value and†

$$\Delta L_{11} = \frac{4}{kA}\left\{k - i\left[\exp\left(-i\frac{k}{2}\right) - 1\right]\right\} \qquad (7.50)$$

$$\Delta L_{12} = \frac{4}{kA}\left\{-i + \frac{k}{2} - \left(\frac{2}{k} + i\right)\left[\exp\left(-i\frac{k}{2}\right) - 1\right]\right\} \qquad (7.51)$$

$$\Delta L_{21} = \frac{4}{kA}\left\{\frac{1}{2}k + \frac{1}{k}\left[\exp\left(-i\frac{k}{2}\right) - 1\right]\right\} \qquad (7.52)$$

$$\Delta L_{22} = \frac{4}{kA}\left\{\frac{1}{k} + \frac{k}{3} - \left(\frac{2i}{k^2} - \frac{1}{k}\right)\left[\exp\left(-i\frac{k}{2}\right) - 1\right]\right\} \qquad (7.53)$$

These remarkably simple formulae are not quite as accurate as results obtained by numerical integration of the two-term solution across the span. Allowing an error of 10 per cent in vector magnitude and 5° in phase angle as in the previous section the formulas above could be used for $A\sqrt{k} > 1.6$ (as compared to $A\sqrt{k} > 1.3$ for the two-term solution). They give quite an improvement over the strip theory which for similar accuracy requires $A\sqrt{k} > 8.5$. The substantial aspect ratio effects present even for large values of $A\sqrt{k}$ show that the strip theory is not a useful approximation at transonic speeds.

7.7. Application to a rectangular control surface

Since there is no upstream influence in linearized transonic flow, the present theory could be used for the calculation of pressures due to the motion of an arbitrarily shaped control surface on any planform with an unswept trailing edge and stream-wise side edges. Only the simplest case of a rectangular control surface will be treated, however. Throughout this section the reference length and reference area used are the control surface chord and area, respectively.

For the modes f_3 and f_4, as given by Eqs. (1.68) and (1.69), Eq. (7.5) gives for $\Phi^{(0)}$ (assuming $W = $ const. for $|y| > \sigma$ as in the preceding section)

$$\Phi^{(0)} = iW \operatorname{sgn}(y - \sigma_1)\int_0^{r_1} H_0^{(2)}(K\eta)\, d\eta \qquad (7.54)$$

where

$$r_1 = |y - \sigma_1|$$

† The formulae for $\Delta L_{12} - \Delta L_{22}$ given in the original publication (Ref. 38) were in error as pointed out to the author by Dr. D. E. Davies of the Royal Aircraft Establishment, Farnborough.

It is convenient to separate out the corresponding strip-theory value $\Phi^{(s)}$. Thus we write

$$\Delta\Phi^{(0)} = \Phi^{(0)} - \Phi^{(s)} = iW \operatorname{sgn}(y - \sigma_1)\left[\int_0^{r_1} H_0^{(2)}(K\eta)\, d\eta - \frac{1}{K}\right] \quad (7.55)$$

The inversion of this result can easily be obtained by aid of Eqs. (7.27) and (7.35). The associated coefficients which are denoted by $\Delta l_{ij}^{(0)}$ ($i, j = 3, 4$) turn out to be given by the result, Eq. (7.38), for $k_{ij}^{(1)}$ in Section 7.5 with subscripts 1 and 2 replaced by 3 and 4, respectively, and with $k_1(x)$ replaced by $k_2(x)$, where

$$k_2(x) = \operatorname{sgn}(y - \sigma_1)\sqrt{\left(\frac{x}{\pi k}\right)}[g_2(v_1) + 1 - i] \quad (7.56)$$

$$g_2 = -2C\left(\frac{v_1^2}{2}\right) + 2iS\left(\frac{v_1^2}{2}\right) + \frac{v_1}{\sqrt{\pi}}\left[Ci\left(\frac{v_1^2}{2}\right) - iSi\left(\frac{v_1^2}{2}\right) + \frac{\pi i}{2}\right] \quad (7.57)$$

and

$$v_1^2 = r_1\sqrt{\frac{k}{x}} \quad (7.58)$$

For the calculation of $\Psi'^{(1)}$ we will assume that $K(\sigma - |\sigma_1|)$ is so large that $\Phi^{(0)}$ can be replaced by its asymptotic value in Eqs. (7.6) and (7.7). From Eq. (7.55) we find, by use of the asymptotic formula for $H_0^{(2)}$,

$$\Delta\Phi^{(0)} \sim -W \operatorname{sgn}(y - \sigma_1)\frac{(1+i)}{(K^{3/2})}\frac{e^{-iKr_1}}{\sqrt{(\pi r_1)}} \quad (7.59)$$

Hence the formulas given in Section 7.5 for $\Psi'^{(1)}$ and $\Psi'^{(2)}$, Eqs. (7.37) and (7.43), could be directly applied so that

$$\Psi'^{(1)}(r) = \frac{i-1}{K}\left[C(Kr) - iS(Kr) - \frac{1}{2} + \frac{i}{2}\right]W$$
$$- \frac{1}{\pi}\int_0^\infty \sqrt{\left[\frac{r(2\varrho + r + d)}{\varrho(\varrho + d)}\right]}\frac{\Delta\Phi^{(0)}(2\varrho + r + d)}{r + \varrho}\, d\varrho \quad (7.60)$$

where $d = \sigma - \sigma_1$ is the distance from the control surface side edge to the wing side edge. For $\bar{\Psi}'^{(1)}$ a similar formula is obtained. The results for the aerodynamic coefficients may thus be written

$$k_{ij}^{(1)} = k_{i-2,j-2}^{(1)} + \Delta k_{i,j}^{(1)} \quad (7.61)$$

where

$$\Delta k_{ij}^{(1)} = -\frac{1}{\pi}\int_0^\infty \sqrt{\left[\frac{r(2\varrho + r + d)}{\varrho(\varrho + d)}\right]}\frac{\Delta k_{ij}^{(0)}(2\varrho + r + d)}{r + \varrho}\, d\varrho \quad (7.62)$$

Comparing Eqs. (7.59) and (7.41) we find that

$$\Delta k_{ij}^{(1)}(r; d) \approx \sqrt{(2)}k_{i-2,j-2}^{(2)}(r; d) \quad (7.63)$$

where the coefficients $k_{i-2,j-2}^{(2)}(r; d)$ for the rigidly oscillating rectangular wing are given by Eq. (7.44) with A replaced by d.

For the calculation of $\overline{\Psi}^{(2)}$ the largest contribution will come from the first term in Eq. (7.60) whereas the second term could be neglected. Hence the only extra coefficients needed for the rectangular control surface in addition to those for rigid-body wing motion in Section 7.5 are $\Delta l_{ij}^{(0)}$ given by Eqs. (7.56)–(7.58).

However, in calculating the coefficients shown in Figs. 7.7 and 7.8 Eq. (7.62) was used, and Eq. (7.63) was only employed as a check.

By suitable superposition of the side-edge effects from the different side edges any configuration of one or several control surfaces may be treated. The approximate Eqs. (7.59)–(7.63) require that $d\sqrt{k}$ is not too small. In particular, if $d\sqrt{k}$ is large and, furthermore, $A_c\sqrt{k}$ is large (A_c is the control surface aspect ratio) the last term in Eq. (7.60) may be neglected and the effect from each side edge treated separately. It is then possible to give approximate formulas for total forces and moments on the control surface (L_{ij} with $i, j = 5, 6$) corresponding to those given in Section 7.6 for the rigidly oscillating rectangular wing. This requires the integral

$$I = \int_0^{A_c} \Delta\Phi^{(0)}(r_1)\, dr_1 \approx \int_0^\infty \Delta\Phi^{(0)}(r_1)\, dr_1 = -\frac{2}{\pi K^2}\, W \qquad (7.64)$$

This differs only by a constant factor from the corresponding result, Eq. (7.47), for the rigidly oscillating rectangular wing. Thus the formulae of Section 7.6, Eqs. (7.49)–(7.53), for the coefficients $\Delta L_{ij}(k; A)$ ($i, j = 1, 2$) may be directly used by setting

$$\Delta L_{ij}(k; A_c) = C\,\Delta L_{i-4, j-4}(k; A_c) \qquad (7.65)$$

where

$$C = \begin{cases} 2/\pi & \text{for an inboard control surface} \\ 1/2(2/\pi + 1) & \text{for an outboard control surface} \\ 1 & \text{for a full-span control surface} \end{cases} \qquad (7.66)$$

In particular we get for the total hinge moment (hinge at $x = 0$, i.e. the leading edge of the control surface)

$$\Delta L_{66} = \frac{4C}{kA_c}\left[\frac{1}{k} + \frac{k}{3} - \left(\frac{2i}{k^2} - \frac{1}{k}\right)\left(\exp\left(-i\frac{k}{2}\right) - 1\right)\right] \qquad (7.67)$$

Examples of numerical applications are shown in Figs. 7.7 and 7.8. Figure 7.7 was prepared in order to demonstrate the effects of finite wing span. The configuration treated is a control surface of aspect ratio three oscillating about its hinge (at the leading edge) with a reduced frequency of 0.2 at $M = 1$. Four different values of the distance, d, from the port wing side edge to the nearest control-surface side edge are considered, namely $d = \infty$, 2, 1 and 0. (The starboard wing side edge is assumed to be situated so far away that its influence on the pressure distribution due to control surface motion could be neglected.)

The coefficient calculated is l_{61}, i.e. the distribution of lift on the control surface and wing due to control surface oscillations about the hinge. For the case $d = \infty$ only $\varphi^{(0)}$ need be calculated. It is seen from Fig. 7.7 that only for very low values of d ($d < 1$) does the presence of the wing side edge have any appreciable influence on the lift distribution on the control surface. For $d\sqrt{k}$ larger than about 0.5 one could therefore calculate forces and moments on the

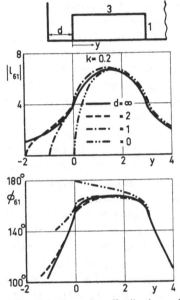

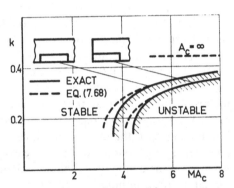

FIG. 7.7. Spanwise distribution of lift due to oscillation about the hinge ($x = 0$) for an $A_e = 3$ rectangular control surface at $M = 1$.

FIG. 7.8. Curves for zero hinge moment damping on rectangular control surfaces at transonic speeds.

control surface with good accuracy by setting $d = \infty$, i.e. by use of Kirchhoff's approximation. It should be pointed out, however, that the curve for $d = 1$ given in Fig. 7.7 must be considered very approximate since for this case the value of $d\sqrt{k}$ is only 0.45, and the approximate formula given for $\Psi'^{(1)}$ above, Eq. (7.60), is certainly not good. Nevertheless the trend shown is believed to be correct.

An aerodynamic design parameter of special interest is the control surface hinge-moment damping. This is given by $(1/k)L_{66}''$ in the present notation (if the hinge is assumed at the leading edge). The sign of L_{66}'' determines whether the control surface is aerodynamically stable to one-degree-of-freedom oscillations about the hinge. A negative value of L_{66}'' means that this oscillation is damped whereas for a positive value undamped oscillations can occur.

According to strip theory L_{66}'' is positive for $k < 0.44$. However, expanding Eq. (7.67) for small values of k gives

$$\Delta L_{66}'' = -\frac{C}{kA_c}\left[1 - \frac{k^2}{16} + 0(k^4)\right] \tag{7.68}$$

Hence the three-dimensional effect is stabilizing, and it turns out that a low enough aspect ratio will make the control surface stable for all frequencies. Curves for zero damping for inboard and outboard control surfaces at transonic Mach numbers are shown in Fig. 7.8. The Mach number dependence is given by the similarity law for unsteady transonic flow (see Section 1.8) according to which generalized forces for Mach number M are obtained by multiplying the results at $M = 1$ for a wing of aspect ratio $A \cdot M$ by M^{-1}. Hence $L_{66}'' = 0$ is only a function of k and MA_c. However, the Mach number dependence is given correctly only when $k/|1 - M| \gg 1$ so that it is probably only valid in a very small region around $M = 1$. Nevertheless it indicates that the likelihood for potential-flow one-degree-of-freedom flutter is larger at low supersonic Mach numbers than at high subsonic ones. The curve given by the solid line in Fig. 7.8 was obtained from numerical integration of l_{66}'' whereas the dotted line was obtained by use of Eq. (7.68). As would be expected from Eqs. (7.68) and (7.66) the outboard control surface is more stable than the inboard one. It is also interesting to notice that there exists a value of A_c below which the control surface is always stable regardless of frequency. The reason for this is that at lower values of k the three-dimensional effects will increase as $A_c\sqrt{k}$ decreases and, hence, will counteract the decreasing strip-theory damping. For a given value of A_c the damping will have a minimum at some value of k. This implies that an increase in hinge stiffness in some circumstances may be detrimental to the damping. However, the minimum damping occurs at such a low value of k (of the order 0.2) that linearized theory is probably no longer valid (see Chapter 1).

CHAPTER 8

THE DELTA WING OF ARBITRARY
ASPECT RATIO

8.1. Introduction

IN THE supersonic case the solution for an oscillating delta wing is not as readily obtained as for a rectangular wing. To be true, it is possible to obtain a formal solution by separation in a special set of co-ordinates (Refs. 56, 21, 19), but the computational difficulties are great and the only solutions that seem to be of practical value are those based on expansion in powers of frequency, like that of Watkins and Berman (Ref. 69).

In the transonic case, however, it is possible, by use of the special co-ordinate transformation given in Section 1.10, to obtain the delta-wing solution from that for a corresponding rectangular wing. This method of solution, which was given originally in Refs. 34 and 39, will be described below.

8.2. Transformation of the boundary value problem to that for a rectangular wing

Consider the problem of an oscillating delta wing with its apex located at the origin and semi-span given by $s = \sigma x$. The boundary value problem is that given by Eqs. (1.88)–(1.90) of Section 1.9. Now we introduce the transformation (1.101), namely

$$X = -1/x$$
$$Y = y/x \qquad (8.1)$$
$$Z = z/x$$

By this the delta wing is transformed to a rectangular wing of span 2σ with its leading edge at $X = -\infty$ and the trailing edge at $X = -1$ (see Fig. 8.1). For φ we introduce the transformation given by Eq. (1.104), namely

$$\varphi = x^{-1} \exp\left\{-i\frac{k}{2}[x + 1/x + (y^2 + z^2)/x]\right\}\omega(X, Y, Z) \qquad (8.2)$$

As shown in Section 1.10 this transformation is such that ω becomes a solution of the transonic equation expressed in X, Y, Z. Hence the boundary value problem for the delta wing is transformed to that for the rectangular wing shown in Fig. 8.1 with the tangency condition

$$\omega Z(X, Y, 0) = v(X, Y) \qquad (8.3)$$

where

$$v(X, Y) = X^{-2} \exp\left[-i\frac{k}{2}(X + 1/X + Y^2/X)\right]w(-1/X, -Y/X) \qquad (8.4)$$

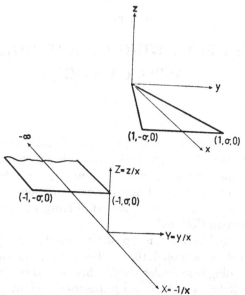

FIG. 8.1. Co-ordinate transformation for the delta wing.

8.3. Series solution for the oscillating delta wing

Using the results of Chapter 7 the solution on the upper surface of the corresponding rectangular wing is given by the following series:

$$\omega = \sum_0^\infty \omega^{(n)} \qquad (8.5)$$

where

$$\omega^{(0)} = -\frac{1}{2\pi} \int_{-\infty}^X dX_1 \int_{-\infty}^\infty dY_1 \frac{v(X_1, Y_1)}{X - X_1} \exp\left\{-i\frac{k}{2}\left[X - X_1 + \frac{(Y - Y_1)^2}{X - X_1}\right]\right\} \qquad (8.6)$$

and

$$\omega^{(n)} = \lambda^{(n)} + \bar{\lambda}^{(n)} \quad \text{for} \quad n \geqslant 1 \qquad (8.7)$$

$$\lambda^{(n)}(X, R) = -\frac{(1 + i)\sqrt{k}}{2\pi^{3/2}} \int_0^\infty \sqrt{\left(\frac{R}{R_1}\right)} dR_1 \int_{-\infty}^X \frac{dX_1}{(X - X_1)^{3/2}}$$

$$\times \exp\left\{-i\frac{k}{2}\left[X - X_1 + \frac{(R + R_1)^2}{X - X_1}\right]\right\}\bar{\lambda}^{(n-1)}(X_1, 2\sigma + R_1) \qquad (8.8)$$

$$\bar{\lambda}^{(n)}(X, \bar{R}) = -\frac{(1 + i)\sqrt{k}}{2\pi^{3/2}} \int_0^\infty \sqrt{\left(\frac{\bar{R}}{\bar{R}_1}\right)} dR_1 \int_{-\infty}^X \frac{dX}{(X - X_1)^{3/2}}$$

$$\times \exp\left\{-i\frac{k}{2}\left[X - X_1 + \frac{(\bar{R} + \bar{R}_1)^2}{X - X_1}\right]\right\}\lambda^{(n-1)}(X_1, 2\sigma + \bar{R}_1) \qquad (8.9)$$

with $\lambda^{(0)}$ and $\bar{\lambda}^{(0)}$ replaced by $\omega^{(0)}$ for $n = 1$.

In Eqs. (8.8) and (8.9) R and $\bar{R}$ denote the distances to the starboard and port side edges, respectively, hence $R = \sigma - Y$ and $\bar{R} = \sigma + Y$ (see Fig. 8.2). We have also introduced the integration variables $R_1 = Y_1 - \sigma$ and $\bar{R}_1 = -\sigma - Y_1$.

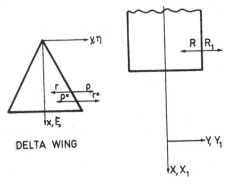

DELTA WING

TRANSFORMED WING

FIG. 8.2. Definitions of co-ordinates.

By use of Eqs. (8.1), (8.2) and (8.3) we may now give the result for the delta wing. In Eq. (8.6) we introduce

$$\varphi^{(0)} = \frac{1}{x} \exp\left[-i\frac{k}{2}(x + 1/x + y^2/x)\right]\omega^{(0)} \tag{8.10}$$

and as integration variables

$$\begin{aligned} X_1 &= -1/\xi \\ Y_1 &= \eta/\xi \end{aligned} \tag{8.11}$$

Since the Jacobian of X_1, Y_1 with respect to ξ, η is ξ^{-3}, the result for $\varphi^{(0)}$ becomes

$$\varphi^{(0)} = -\frac{1}{2\pi}\int_0^x \frac{d\xi}{x-\xi}\int_{-\infty}^\infty w(\xi,\eta)\exp\left\{-i\frac{k}{2}\left[x - \xi + \frac{(y-\eta)^2}{x-\xi}\right]\right\}d\eta \tag{8.12}$$

This could of course have been obtained more directly but the identical forms of Eqs. (8.6) and (8.12) serve as a check on the transformation. Since w is given only on the wing it is necessary to assume some arbitrary distribution of w outside the wing.

When transforming Eqs. (8.8) and (8.9) we set

$$\psi^{(n)} = \frac{1}{x}\exp\left\{-i\frac{k}{2}(x + 1/x + y^2/x)\right\}\lambda^{(n)}$$

$$\bar{\psi}^{(n)} = \frac{1}{x}\exp\left\{-i\frac{k}{2}(x + 1/x + y^2/x)\right\}\bar{\lambda}^{(n)} \tag{8.13}$$

We also introduce

$$R = r/x \qquad \bar{R} = \bar{r}/x$$
$$R_1 = \varrho/\xi \qquad \bar{R}_1 = \bar{\varrho}/\xi \tag{8.14}$$

$r, \bar{r}, \varrho$ and $\bar{\varrho}$ are thus related to y and η as follows (see Fig. 8.2):

$$r = s(x) - y$$
$$\bar{r} = s(x) + y$$
$$\varrho = \eta - s(\xi)$$
$$\bar{\varrho} = -\eta - s(\xi) \tag{8.15}$$

With these substitutions the results for $\psi^{(n)}$ and $\bar{\psi}^{(n)}$ become

$$\psi^{(n)}(x, r) = -\frac{1}{\pi} \int_0^x d\xi \int_0^\infty d\varrho \; h(x - \xi, r, \varrho; \sigma) \bar{\psi}^{(n-1)}(\xi, 2s(\xi) + \varrho) \tag{8.16}$$

$$\bar{\psi}^{(n)}(x, \bar{r}) = -\frac{1}{\pi} \int_x^0 d\xi \int_0^\infty d\bar{\varrho} \; \bar{h}(x - \xi, \bar{r}, \bar{\varrho}; \sigma) \psi^{(n-1)}(\xi, 2s(\xi) + \bar{\varrho}) \tag{8.17}$$

where

$$h(x, r, \varrho; \sigma) = \bar{h}(x, r, \varrho; \sigma) = \frac{1 + i}{2\sqrt{\pi}} \sqrt{\left(\frac{r}{\varrho}\right)} \frac{\sqrt{(k)}}{x^{3/2}}$$

$$\times \exp\left\{-i\frac{k}{2}\left[(1 + \sigma^2)x - 2\sigma(r + \varrho) + \frac{(r +)\varrho^2}{x}\right]\right\} \tag{8.18}$$

The simple relationship to the corresponding kernel g for the rectangular wing should be noted. This kernel is obtained by setting $\sigma = 0$ in Eq. (8.18). Hence

$$h = g \exp\left\{-i\frac{k}{2}[\sigma^2 x - 2\sigma(r + \varrho)]\right\} \tag{8.19}$$

The complete solution is then given by the series

$$\varphi = \sum_0^\infty \varphi^{(n)} \tag{8.20}$$

where

$$\varphi^{(n)} = \psi^{(n)} + \bar{\psi}^{(n)} \quad \text{for} \quad n \geqslant 1 \tag{8.21}$$

and $\varphi^{(0)}$ is given by Eq. (8.12).

8.4. Convergence of the series solution

The convergence of the series solution for the rectangular wing was proved in Ref. 37. It was shown that the series converges for all non-zero values of σ

and k provided the Fourier transform of $\omega^{(0)}$ fulfills certain conditions. Let the Fourier transform of $\omega^{(n)}$ with respect to X be denoted as follows:

$$\Omega^{(n)} = \frac{1}{\sqrt{(2\pi)}} \int_{-\infty}^{\infty} \omega^{(n)} e^{-iuX} \, dX \tag{8.22}$$

Then the series $\sum_{0}^{\infty} \omega^{(n)}$ converges for all X and Y on the wing, provided $\Omega^{(0)}$ on the Y-axis is such that the integral

$$\int_{0}^{\infty} \frac{|\Omega^{(0)}(\sigma + R_1, +0; u)| \, dR_1}{\sqrt{(R_1)(2\sigma + R_1)}} \tag{8.23}$$

exists for all u in the lower half-plane. The critical point for convergence of the Fourier integral is $X = -\infty$. (Since there is no upstream influence the potential for $X > -1$ can be set arbitrarily equal to zero without affecting the value on the wing.) If for $\varphi^{(0)}$ near the apex is taken the two-dimensional result we find that, as $x \to 0$,

$$\varphi^{(0)} \sim x^{1/2} \tag{8.24}$$

Hence from Eq. (8.10) it follows that, as $X \to -\infty$,

$$\omega^{(0)} \sim (-X)^{-3/2} \tag{8.25}$$

and the Fourier integral can thus always be made to converge for all u and Y if $v(X, Y)$ is suitably chosen outside the wing. Consequently the integral (8.23) will also converge and the series solution is thus convergent for all points on the wing. However, for the *loading* near the apex, the series will not converge, as is evident from the following simple consideration.

Consider the loading in the region near the apex for $0 \leqslant x \leqslant \varepsilon$, where ε is small compared to unity. Since there is no upstream influence the loading can be found by considering the solution for a delta wing with a root chord of ε. We now introduce the following transformations:

$$x_1 = x/\varepsilon$$

$$y_1 = y/\varepsilon$$

$$z_1 = z/\varepsilon \tag{8.26}$$

$$\varphi_1 = \varphi/\varepsilon$$

Then the problem becomes that for a delta wing of unit chord oscillating at the reduced frequency of $k_1 = \varepsilon k$ with the tangency condition

$$\varphi_{1z_1} = w(\varepsilon x_1, \varepsilon y_1) \tag{8.27}$$

where $w(x, y)$ is given by Eq. (8.28). As $\varepsilon \to 0$, $k_1 \to 0$ and the series solution will not converge. Hence the solution given by Eqs. (8.12)–(8.21) is such that

$$\varphi = \sum_0^\infty \varphi^{(n)}$$

converges for all x, y on the wing but that

$$\frac{1}{x}\varphi = \sum_0^\infty \frac{1}{x}\varphi^{(n)} \tag{8.28}$$

does not converge for $x = y = 0$. Since, for $y = 0$, $\lim\limits_{x \to 0} (\varphi/x) = \varphi_x(0, 0)$, this means that the loading at the apex cannot be found from the present solution.

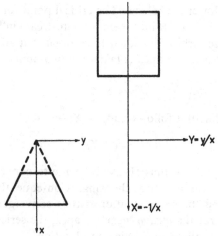

Fig. 8.3. Transformation of the cut-off delta wing.

However, this is not a serious restriction from a practical point of view and in any case the correct loading at the apex can be found from slender-wing theory which gives

$$\varphi(x, y, +0) = -w(0, 0)\sqrt{(s^2 - y^2)} \tag{8.29}$$

as $x \to 0$.

It is interesting to note that the present solution is valid also for a delta wing with a cut-off apex (see Fig. 8.3). In this case the corresponding rectangular wing has a finite chord and the difficulty in connection with the apex will not occur.

8.5. Approximate solution for low aspect ratios

The oscillating low aspect ratio delta wing was treated in Chapter 3 by use of the Adams and Sears (Ref. 1) iteration method. With the present transformation an independent low aspect ratio solution can be deduced by employing the solution given in Chapter 4 for a low aspect ratio rectangular wing. This was obtained as a Glauert series as follows:

$$\omega = \frac{2\sigma}{\pi} \sum_{n=1}^{\infty} \omega_n(x) \sin n\theta \tag{8.30}$$

where

$$\psi = \cos^{-1}(Y/\sigma) \tag{8.31}$$

$$\omega_1 = \omega_{11}(v_1) + \omega_{13}(v_3) + 0(\sigma^4 k^2) \tag{8.32}$$

$$\omega_n = \omega_{n,n-2}(v_{n-2}) + \omega_{n,n}(v_n) + \omega_{n,n+2}(v_{n+2}) + 0(\sigma^4 k^2) \quad \text{for} \quad n \geqslant 3 \tag{8.33}$$

$$v_n = \int_0^{\pi} \sin \theta \sin n\theta \, v \, d\theta \tag{8.34}$$

In Chapter 4 terms of order $\sigma^4 k^2$ were also included in the solution but here only terms of order $\sigma^2 k$ are retained for simplicity. To this order the functions ω_{11} and ω_{13} are given by

$$\omega_{11} = -\left[1 - \frac{ik\sigma^2}{4}\left(\frac{3}{2} - v\right)D\right]v_1(X)$$
$$+ \frac{ik\sigma^2}{4} D^2 \int_{-\infty}^{X} \exp\left[-i\frac{k}{2}(X - X_1)\right] \ln(X - X_1) \, v_1(X_1) \, dX_1 + E_{11} \tag{8.35}$$

$$\omega_{13} = -\frac{ik\sigma^2}{32}\left(\frac{4}{3} - v\right)D \, v_3(X) - \frac{ik\sigma^2}{24} D^2 \int_{-\infty}^{X} \exp\left[-i\frac{k}{2}(X - X_1)\right]$$
$$\times \ln(X - X_1) \, v_3(X_1) \, dX_1 + E_{13} \tag{8.36}$$

where

$$v = \gamma - \ln \frac{8}{k\sigma^2} + \frac{\pi i}{2}$$

$$D = \frac{\partial}{\partial X} + i\frac{k}{2}$$

and where E_{11} and E_{13} stand for the exponential terms

$$E_{11} = a_{11}\left\{\frac{2i}{k\sigma^2} \int_{-\infty}^{X} \exp\left[\frac{2}{k\sigma^2} B(X - X_1)\right] v_1(X_1) \, dX_1 \right.$$
$$\left. + \sum_{m=0}^{\infty} B^{-m-1}\left(-i\frac{k\sigma^2}{2} \frac{\partial}{\partial X}\right)^m v_1(X)\right\} \tag{8.37}$$

$$E_{13} = a_{13}\left\{\frac{2i}{k\sigma^2} \int_{-\infty}^{\infty} \exp\left[\frac{2i}{k\sigma^2} B(X - X_1)\right] v_3(X_1) \, dX_1 \right.$$
$$\left. + \sum_{m=0}^{\infty} B^{-m-1}\left(-i\frac{k\sigma^2}{2} \frac{\partial}{\partial X}\right)^m v_3(X)\right\} \tag{8.38}$$

where

$$a_{11} = 0.088 + 0.486i$$
$$a_{13} = 0.021 - 0.017i \tag{8.39}$$
$$B = 0.192 + 0.381i - k^2\sigma^2/4$$

The series in E_{11} and E_{13} (which are actually semi-convergent) serve to cancel the upper limit in the integrals. Performing a number of integrations by parts gives

$$E_{11} = \frac{a_{11}}{B}\sum_\mu\sum_\gamma\left(\frac{k\sigma^2}{2iB}\right)^\nu \Delta v_1^{(\nu)}(X_\mu)\exp\left[\frac{2iB}{k\sigma^2}(X-X_\mu)\right] \tag{8.40}$$

$$E_{13} = \frac{a_{13}}{B}\sum_\mu\sum_\nu\left(\frac{k\sigma^2}{2iB}\right)^\nu \Delta v_3^{(\nu)}(X_\mu)\exp\left[\frac{2iB}{k\sigma^2}(X-X_\mu)\right] \tag{8.41}$$

where $\Delta v_n^{(\nu)}(X_\mu)$ is the jump in $\partial^\nu v_n/\partial X^\nu$ at $X = X_\mu$.

For ω_{31} the same formula (8.36) as for ω_{13} is valid except that v_3 should be replaced by v_1. Remaining terms in the series solution do not contain any exponential terms. The solution for the triangular wing may now be obtained with the aid of Eqs. (8.2) and (8.3). Upon expanding the exponential terms in these equations to the proper order in σ,

$$\frac{1}{x}\exp\left[-i\frac{k}{2}(x+1/x+y^2/x)\right] = \frac{1}{x}\exp\left[-i\frac{k}{2}(x+1/x)\right]\left[1-i\frac{ky^2}{2x}+0(k^2\sigma^4)\right] \tag{8.42}$$

it turns out that the results will be identical with that obtained in Chapter 3 with exception of the exponentials terms. If $w(x,y)$ is a smooth function in x the exponential terms will be zero and we have thus rederived the result of Chapter 3. Equations (8.2), (8.3), (8.40) and (8.41), however, give also the additional terms which should be added when the distribution is not smooth. Thus, for example, a jump $\Delta w(x_0,y)$ in $w(x,y)$ at $x = x_0$, where $s = s_0$, gives the following term $\Delta\varphi$ to be added to the result for φ obtained by Adams–Sears' method:

$$\Delta\varphi = \frac{2s}{\pi B}\exp\left[-i\frac{k}{2}(x-x_0+1/x-1/x_0)\right]\exp\left[\frac{2iB}{k}\left(\frac{x_0}{s_0^2}-\frac{x}{s^2}\right)\right]\left(\frac{s_0}{s}\right)^2$$

$$\times[(a_{11}\Delta w_1 + a_{13}\Delta w_3)\sin\theta + a_{13}\Delta w_1\sin 3\theta + 0(k\sigma^2)] \tag{8.43}$$

where now

$$\theta = \cos^{-1}(y/s)$$

and where

$$\Delta w_n = \int_0^\pi \sin n\theta\,\sin\theta\Delta w\,d\theta \tag{8.44}$$

CHAPTER 9

WINGS OF GENERAL PLANFORMS

9.1. Introduction

IT WAS seen in Chapters 7 and 8 that the solution for rectangular and delta wings could be obtained by a process of successive cancelation of lift outside the wing edges. Such a process is of course not restricted to these planforms, but could be used for wings of planforms consisting of polygons that are convex outwards. For concave planforms the process has to be supplemented by upwash-cancelations. Applications of the lift- and upwash-cancelation techniques for steady and quasi-steady supersonic flow have been discussed by Behrbohm (Ref. 4). The applications to unsteady transonic flow were originally given in Ref. 39. In the present chapter the so-called kernel-function method for numerical solution of the lifting-surface problem is also briefly described.

9.2. Convex polygonal planforms

The solution for the delta wing, and in particular that for the cut-off delta (Fig. 8.3), can be used to build up the general solution for wings of planforms consisting of polygons that are convex outwards. Such a planform is shown in Fig. 9.1.

The solution for the original planform S_0 is constructed from solutions for the planforms S_1, S_2 and S_3 as shown in the figure. The planform S_1 is the cut-off delta that coincides with the original planform up to the line $B\bar{B}$. For S_1 the solution is given by the series in Section 8.3. Originally, w is not given in the regions BC_1DC and $\bar{B}\bar{C}_1\bar{D}\bar{C}$ so that some suitable distributions in these have to be assumed. The solution for S_0 is now obtained up to the line $B\bar{B}$.

The solution φ_1 for S_1 is now taken as the starting solution $\varphi_2^{(0)}$ for S_2. This solution fulfills all boundary conditions except that it is not zero in the regions BC_1C_2 and $\bar{B}\bar{C}_2\bar{C}_1$. $\varphi_2^{(0)} = \varphi_1$ is therefore canceled in these regions by aid of Eqs. (8.16) and (8.17). The solution for S_0 is now completed up to the line $C\bar{C}$.

This solution φ_2, finally, satisfies all boundary conditions for S_1, except that it does not give a zero pressure in the wake. In canceling the wake pressure it is more convenient, as was done in Section 2.4, to work directly with the pressure coefficient:†

$$C_p = -2(\varphi_x + ik\varphi)e^{ikt} \tag{9.1}$$

† In Ref. 39 a solution for the wake influence based on the velocity potential was presented. However, this solution is actually incorrect since it does not satisfy the Kutta condition.

99

which satisfies the same differential equation (1.88) as φ. Therefore, since in addition

$$C_{pz}(x, y, 0) = -2(w_x + ikw)e^{ikt} \qquad (9.2)$$

and the required pressure cancelation function thus should have a zero normal derivative on the wing we can directly make use of the kernel h, Eq. (8.18),

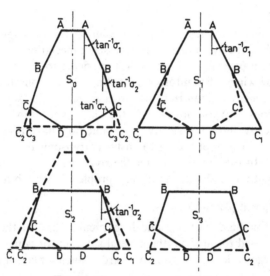

FIG. 9.1. Convex polygonal planform.

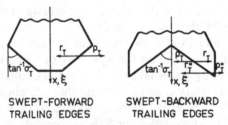

SWEPT-FORWARD   SWEPT-BACKWARD
TRAILING EDGES   TRAILING EDGES

FIG. 9.2. Definition of co-ordinates.

derived for the potential function. Hence, denoting the pressure coefficient in the wake associated with φ_2 by C_{p2} and the pressure coefficient that arises from the cancelation of this by C_{pw} we find

$$C_{pw} = \sum_{1}^{\infty} (C_{pw}^{(n)} + \bar{C}_{pw}^{(n)}) \qquad (9.3)$$

where

$$C_{pw}^{(n)} = -\frac{1}{\pi}\int_{x_0}^{x} d\xi \int_0^{\infty} h_T \bar{C}_{pw}^{(n-1)}\, d\rho_T \qquad (9.4)$$

$$\bar{C}_{pw}^{(n)} = -\frac{1}{\pi}\int_{x_0}^{x} d\xi \int_0^{\infty} h_T C_{pw}^{(n-1)}\, d\bar{\rho}_T \qquad (9.5)$$

and the variables r_T, ρ_T, etc., referring to the trailing edge are defined in Fig. 9.2. For $n = 0$, $C_{pw}^{(n)}$ and $\bar{C}_{pw}^{(n)}$ should be replaced by C_{p2} in Eqs. (9.4) and (9.5). The series (9.3) converges for all non-zero values of k and local wing span.

A difficulty which has so far been neglected, is that C_{p2} has a square-root singularity along CC_2 and $\bar{C}\bar{C}_2$. A separate investigation shows, however, that no error will be introduced as long as the singularity is integrable.

9.3. Calculation of the upwash outside the wing edges

For the treatment of concave polygonal planforms it is necessary to know the upwash outside the leading edges of the basic planform, the cut-off delta. This can be found in several different ways. The conceptually simplest one is to express the solution for φ as a distribution of doublets on the wing. Then

$$\varphi_z = -\frac{1}{2\pi}\frac{\partial^2}{\partial z^2}\int_0^x \frac{d\xi}{x-\xi}\int_{-s(\xi)}^{s(\xi)} \exp\left\{-i\frac{k}{2}\left[(x-\xi) + \frac{(y-\eta)^2 + z^2}{x-\xi}\right]\right\}$$
$$\times\, \varphi(\xi, \eta, +0)\, d\eta \qquad (9.6)$$

Outside the wing $\varphi(x, y, 0) = 0$ and we may write the result for $\varphi_z(x, y, 0) = w^*(x, y)$ as follows:

$$w^* = \frac{1}{\pi}\int_{-s(x)}^{s(x)} \frac{d\eta}{(y-\eta)^2} \exp\left[-i\frac{k}{2}\frac{(y-\eta)^2}{x-x_L}\right]\varphi(x, \eta)$$
$$+ \frac{ik}{2\pi}\int_0^x \frac{d\xi}{(x-\xi)^2}\int_{-s(\xi)}^{s(\xi)}\left[\varphi(\xi, \eta)\exp\left[-i\frac{k}{2}(x-\xi)\right] - \varphi(x, \eta)\right]$$
$$\exp\left[-i\frac{k}{2}\frac{(y-\eta)^2}{x-\xi}\right]d\eta \qquad (9.7)$$

where $x_L(\eta)$ is the abscissa of the leading edge at $y = \eta$. The term added and subtracted serves the purpose of making the integral converge at $x = \xi$.

A second method is to try to express w^* directly in terms of w on the wing. Consider first the problem for the starboard side edge of a rectangular wing. It was found in Section 7.2 that a solution of the wave equation (1.93) in the Fourier transformed plane could be given in a polar co-ordinate system (see Fig. 9.3) as follows:

$$\Phi = \frac{1}{\pi}\int_0^{\infty} G(r, \rho, \theta)F(\rho)\, d\rho \qquad (9.8)$$

where G is the Green's function defined in Section 7.3. G is such that $G_\theta = 0$ for $\theta = \pi$ and that for $\theta = 0$ G behaves like a delta function, i.e.

$$\Phi(r) = \frac{1}{\pi}\int_0^\infty G(r, \rho, 0) F(\rho)\, d\rho \doteq F(r) \tag{9.9}$$

Now we construct a second solution Φ^* by setting

$$\Phi^* = \frac{1}{\pi}\int_0^\infty G^*(r, \rho^*, \theta)F^*(\rho^*)\, d\rho^* \tag{9.10}$$

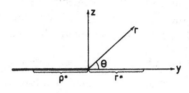

Fig. 9.3.

where the integration variable ρ^* now is measured inboard from the edge (see Fig. 9.2), and where

$$G^* = \int_0^\theta G(r, \rho^*, \pi - \theta_1)\, d\theta_1 \tag{9.11}$$

Hence $\Phi^* = 0$ for $\theta = 0$. From Eqs. (9.8)–(9.11) it follows that

$$\Phi_\theta^*(r, \pi) = F^*(r) \tag{9.12}$$

Now $\Phi_\theta^*(\rho^*, \pi) = -\rho^* W(\rho^*)$ and $\Phi_\theta^*(r, 0) = r^* W^*(r^*)$, where r^* is measured outboard from the edge (see Fig. 9.2). Hence

$$W^*(r^*) = -\frac{1}{\pi}\int_0^\infty \frac{\rho^*}{r^*} G(r^*, \rho^*, \pi)W(\rho^*)\, d\rho^* \tag{9.13}$$

Or, introducing G from Eq. (7.26),

$$W^*(r^*) = -\frac{1}{\pi}\int_0^\infty \sqrt{\left(\frac{\rho^*}{r^*}\right)}\frac{\exp\left[-iK(r^* + \rho^*)\right]}{r^* + \rho^*} W(\rho^*)\, d\rho^* \tag{9.14}$$

Upon inversion we obtain

$$w^* = -\frac{1}{\pi}\int_0^x d\xi \int_0^\infty h^*(x - \xi, r^*, \rho^*)\, w(\xi, \rho^*)\, d\rho^* \tag{9.15}$$

where

$$h^*(x, r^*, \rho^*) = \frac{1 + i}{2\sqrt{\pi}} \sqrt{\left(\frac{\rho^*}{r^*}\right)}\frac{\sqrt{k}}{x^{3/2}}\exp\left\{-i\frac{k}{2}\left[x + \frac{(r^* + \rho^*)^2}{x}\right]\right\} \tag{9.16}$$

The extension to the case of a delta-wing leading edge is easily obtained by the transformations given in Section 1.10. The result turns out to be that given by Eq. (9.16), except that $h^*(x - \xi, r^*, \rho^*) = h^*(x - \xi, r^*, \rho^*; 0)$ should be replaced by $h^*(x - \xi, r^*, \rho^*; \sigma)$ where

$$h^*(x, r^*, \rho^*; \sigma) = h^*(x, r^*, \rho^*; 0) \exp\left\{-i\frac{k}{2}\left[\sigma^2 x - 2\sigma(r^* + \rho^*)\right]\right\} \quad (9.17)$$

Note that the integration in Eq. (9.15) is to be extended also outside the port edge, i.e. where w is not known beforehand. Let $\bar{w}^{*(0)}$ be the w-distribution that is assumed outside the port edge for the calculation of $\varphi^{(0)}$ and $w^{*(0)}$ the corresponding distribution outside the starboard edge. Consider first the potential function $\varphi^{*(1)} = \varphi^{(0)} + \psi^{(1)}$ (for definitions of $\varphi^{(0)}$ and $\psi^{(1)}$ see Chapter 8). This is zero outside the starboard edge as required and, since $\psi^{(1)}_z = 0$ to the left of the starboard edge, $\varphi^{*(1)}_z = w$ on the wing and $\varphi^{*(1)}_z = \bar{w}^{*(0)}$ to the left of the port edge. Application of Eq. (9.15) then gives for $\varphi^{*(1)}_z$ outside the starboard edge

$$\varphi^{*(1)}_z = w^{*(1)} = -\frac{1}{\pi}\int_0^x d\xi \int_0^{2s} h^* w \, d\rho^* - \frac{1}{\pi}\int_0^x d\xi \int_{2s}^{\infty} h^* \bar{w}^{*(0)} \, d\rho^*$$
$$(9.18)$$

Similarly, by considering the potential function $\varphi^{*(2)} = \bar{\psi}^{(1)} + \psi^{(2)}$ we get for $\varphi^{*(2)}_z = w^{*(2)}$ outside the edge

$$w^{*(2)} = -\frac{1}{\pi}\int_0^x d\xi \int_{2s}^{\infty} (\bar{w}^{*(1)} - \bar{w}^{*(0)})h^* \, d\rho^* \quad (9.19)$$

where $\bar{w}^{*(1)}$ is obtained from Eq. (9.18) by replacing unbarred quantities by barred ones and vice versa. By continuing in this manner we obtain for $n > 2$

$$w^{*(n)} = -\frac{1}{\pi}\int_0^x d\xi \int_{2s}^{\infty} \bar{w}^{*(n-1)} h^* \, d\rho^* \quad (9.20)$$

This leads to the following series expression for w^* and $\bar{w}^*$

$$w^* = \sum_{n=0}^{\infty} w^{*(n)} \quad (9.21)$$

$$\bar{w}^* = \sum_{n=0}^{\infty} \bar{w}^{*(n)} \quad (9.22)$$

where the $\bar{w}^{*(n)}$: s are obtained by replacing all barred quantities with unbarred ones, and vice versa.

9.4. Concave polygonal planforms

A typical concave polygonal wing planform is shown in Fig. 9.4. The solution for the original planform S_0 is constructed from solutions for the planforms S_1, S_2 and S_3, shown in the figure. The solution for the cut-off delta S_1 is first obtained by aid of the formulas given in Chapter 8. The solution for S_0 is now obtained up to the line $B\bar{B}$.

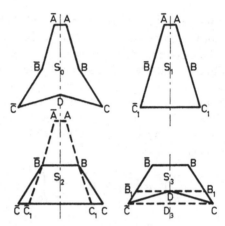

FIG. 9.4. Concave polygonal planform.

The solution for S_1', φ_1, is taken as the starting solution for S_2. This solution fulfills all boundary conditions except that the normal derivative distributions on BCC_1 and $\bar{B}\bar{C}_1\bar{C}$ will be incorrect since these regions are outside the leading edges of S_1. We denote the normal derivative distribution outside the right edge by w_1^* and that outside the left edge by $\bar{w}_1^*$. In Section 9.3 the calculation of these distributions was considered. It was found that they can be expressed in form of the series,

$$w_1^* = \sum_0^\infty w_1^{*(n)} \tag{9.23}$$

$$\bar{w}_1^* = \sum_0^\infty \bar{w}_1^{*(n)} \tag{9.24}$$

where the individual terms are defined by Eqs. (9.18)–(9.22).

The series for w_1^* and $\bar{w}_1^*$ converge under the same conditions as the series for φ_1. Having w_1^* and $\bar{w}_1^*$ the problem for the planform S_2 is to find a solution $\Delta\varphi_2$ with $\Delta\varphi_{2z} = w - w_1^*$ on BCC_1 and $\Delta\varphi_{2z} = w - \bar{w}_1^*$ on $\bar{B}\bar{C}_1\bar{C}$ but with $\Delta\varphi_{2z} = 0$ on the rest of the wing. This solution is readily obtained as before, and $\varphi_2 = \varphi_1 + \Delta\varphi_2$ will then fulfill all boundary conditions for the original planform, except that the pressure will not be zero in the wake region $C\bar{C}D$.

The calculation of the wake influence in this case is considerably more complicated than for the convex planform. The reason is that, when the pressure in the wake is canceled, normal velocities will be introduced on the swallow-tail regions of the wing behind $x = x_D$. These velocities have in turn to be canceled and this cancellation will introduce pressures both outside the leading edges and in the wake region, etc.

Like in the case of the convex planform, it is convenient to work with the pressure coefficient directly. Let the starting solution $\varphi_3^{(0)}$, for the wing S_3 be the solution φ_2 for the wing S_2 and its associated pressure coefficient $C_{p3}^{(0)} = C_{p2}$. To cancel $C_{p3}^{(0)}$ in the wake regions we add the pressure coefficients $C_{pw}^{(1)}$ and $\bar{C}_{pw}^{(1)}$ defined by

$$C_{pw}^{(1)} = -\frac{1}{\pi} \int_{x_D}^{x} d\xi \int_0^\tau h_T C_{p3}^{(0)} \, d\rho_T \tag{9.25}$$

$$\bar{C}_{pw}^{(1)} = -\frac{1}{\pi} \int_{x_D}^{x} d\xi \int_0^\tau \bar{h}_T C_{p3}^{(0)} \, d\bar{\rho}_T \tag{9.26}$$

The variables r_T, ρ_T, etc., are defined in Fig. 9.2 and h_T and $\bar{h}_T$ as in Eqs. (9.4) and (9.5). τ is the local trailing-edge semi-span at the station x. Note that we have chosen $C_{pw}^{(1)}$ and $\bar{C}_{pw}^{(1)}$ so that they cancel $C_{p3}^{(0)}$ in the wake for $y > 0$ and $y < 0$, respectively. Other choices are of course also possible, but the present one seems to be the most practical.

If the trailing edges are far separated from each other and from the leading edges, or the frequency is high, the pressure coefficient given by

$$C_{p3} = C_{p3}^{(0)} + C_{p3}^{(1)} \tag{9.27}$$

where

$$C_{p3}^{(1)} = C_{pw}^{(1)} + \bar{C}_{pw}^{(1)} \tag{9.28}$$

would give a good approximation to the true solution. This will be the case if $\tau\sqrt{k}$ and $(s - \tau)\sqrt{k}$ are large for all x which requires a trailing-edge shape like the one shown in Fig. 9.5. However, for finite values of these parameters

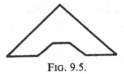

FIG. 9.5.

$(\partial/\partial z)\, C_{p3}^{(1)}$ will not be zero on DB_1C and $D\bar{C}\bar{B}_1$ as is required if the normal velocity on the wing should be unaffected. (Here we have assumed that C_{p2} is finite in the wake so that $C_{p3}^{(1)}$ will have no singularities.) Furthermore, $C_{p3}^{(1)}$ will not be zero outside the leading edges. For the successive-cancellation process it is convenient to introduce the notation

$$\left[\frac{\partial}{\partial z} C_p\right]_{z=0} = v(x, y, t) \tag{9.29}$$

with associated sub- and superscript as for C_p. In calculating $v_3^{(1)}$ we remember that on the starboard swallow-tail region DB_1C, there is no contribution from $C_{pw}^{(1)}$ but only from $\bar{C}_{pw}^{(1)}$ by definition. Since $\bar{C}_{pw}^{(1)} = -C_{p3}^{(0)}$ in the port wake region $DD_3\bar{C}$ but zero for $y > 0$ we may use Eq. (9.7) which gives†

$$v_3^{(1)} = \frac{1}{\pi} \int_{-\infty}^{0} \frac{d\eta}{(y-\eta)^2} \exp\left[-i\frac{k(y-\eta)^2}{2x-x_D}\right] \bar{C}_{pw}^{(1)}(x,\eta)\, d\eta$$

$$+ \frac{ik}{2\pi} \int_{x_D}^{x} \frac{d\xi}{(x-\xi)^2} \int_{-\infty}^{0} \left[\bar{C}_{pw}^{(1)}(\xi,\eta) \exp\left[-i\frac{k}{2}(x-\xi)\right] - \bar{C}_{pw}^{(1)}(x,\eta)\right]$$

$$\times \exp\left[-i\frac{k}{2}\frac{(y-\eta)^2}{x-\xi}\right] d\eta \quad (9.30)$$

in the region DB_1C. Note that $y \neq \eta$ so that $v_3^{(1)}$ is finite. Similarly, in the port swallow-tail region $DC\bar{B}_1$ the normal derivative due to $C_{pw}^{(1)}$ is

$$\bar{v}_3^{(1)} = \frac{1}{\pi} \int_{0}^{\infty} \frac{d\eta}{(y-\eta)^2} \exp\left[-i\frac{k}{2}\frac{(y-\eta)^2}{x-x_D}\right] C_{pw}^{(1)}(x,\eta)\, d\eta$$

$$+ \frac{ik}{2\pi} \int_{x_D}^{x} \frac{d\xi}{(x-\xi)^2} \int_{0}^{\infty} \left[C_{pw}^{(1)}(\xi,\eta) \exp\left[-i\frac{k}{2}(x-\xi)\right] - C_{pw}^{(1)}(x,\eta)\right]$$

$$\times \exp\left[-i\frac{k}{2}\frac{(y-\eta)^2}{x-\xi}\right] d\eta \quad (9.31)$$

We now cancel $v_3^{(1)}$ and $\bar{v}_3^{(1)}$ by adding the pressure coefficient $\Delta C_p^{(2)}$ defined by

$$\Delta C_p^{(2)} = \frac{1}{\pi} \int_{x_D}^{x} \frac{d\xi}{x-\xi} \left[\int_{-\infty}^{-\tau} \bar{v}_3^{(1)} \exp\left\{-i\frac{k}{2}\left[x-\xi+\frac{(y-\eta)^2}{x-\xi}\right]\right\} d\eta\right.$$

$$+ \int_{\tau}^{\infty} v_3^{(1)} \exp\left\{-i\frac{k}{2}\left[x-\xi+\frac{(y-\eta)^2}{x-\xi}\right]\right\} d\eta \quad (9.32)$$

To cancel the pressure outside the leading edges it is necessary to first calculate the velocity potential which is given by the formula

$$\varphi_3^{(1)} = -\frac{1}{2} e^{-ikt} \int_{x_D}^{x} C_{p3}^{(1)}(u, y, 0, t-x+u)\, du \quad (9.33)$$

The reason for this is that the correct leading-edge pressure singularity cannot otherwise be found.‡ The cancellation of $\varphi_3^{(1)}$ is readily achieved as above by use of the kernels h and $\bar{h}$ associated with the leading edges. Let the cancellation

† In Ref. 39 an unnecessarily complicated method was proposed for the corresponding calculation.
‡ If the wing is bordered by streamwise side edges for $x > x_p$, instead of by oblique leading edges, the pressure can be canceled directly and the velocity potential need not be calculated.

potentials be denoted by $\psi_3^{(2)}$ and $\bar{\psi}_3^{(2)}$ and the associated pressure coefficients by $C_{pL}^{(2)}$ and $\bar{C}_{pL}^{(2)}$, respectively. By adding all these contributions we thus arrive to the next higher approximation for the pressure in the region influenced by the wake

$$C_{p3} = C_{p3}^{(0)} + C_{p3}^{(1)} + C_{p3}^{(2)} \tag{9.34}$$

where

$$C_{p3}^{(2)} = \Delta C_{p3}^{(2)} + C_{pL}^{(2)} + \bar{C}_{pL}^{(2)} \tag{9.35}$$

Continuing the iteration in this manner we obtain an infinite series whose terms may be written

$$C_{p3}^{(n)} = C_{pw}^{(n)} + \bar{C}_{pw}^{(n)} + C_{pL}^{(n)} + \bar{C}_{pL}^{(n)} + \Delta C_{p3}^{(n)} \tag{9.36}$$

where $C_{pw}^{(n)}$ and $\bar{C}_{pw}^{(n)}$ cancel $C_{p3}^{(n-1)}$ in the wake; $C_{pL}^{(n)}$ and $\bar{C}_{pL}^{(n)}$ arise from canceling the velocity potential $\varphi_3^{(n-1)}$ associated with $C_{p3}^{(n-1)}$ outside the leading edges; and $\Delta C_{p3}^{(n)}$ cancels the normal derivative of $\bar{C}_{p3}^{(n-1)}$ on the wing. Note that, because of the way the iteration was started, $C_{pL}^{(1)} = \bar{C}_{pL}^{(1)} = \Delta C_p^{(1)} = 0$ and $C_{pw}^{(2)} = \bar{C}_{pw}^{(2)} = 0$. $C_{pw}^{(n)}$ and $\bar{C}_{pw}^{(n)}$ can be found from Eqs. (9.25) and (9.26), and $\Delta C_p^{(n)}$ from Eqs. (9.30)–(9.32).

The series converges for all non-zero values of k and wing span, except at the corners D, C and $\bar{C}$, where the situation will be similar to that at the apex of a delta wing as discussed in Section 8.4.

The complete wake pressure cancellation for a wing with swept-backward trailing edges is apparently very cumbersome since four different types of integrals need be evaluated. In many practical cases, however, it would probably suffice to calculate only the first-order wake influence which is given solely by Eqs. (9.25) and (9.26).

9.5. The kernel function method

The only other method available in the literature which can be used for the evaluation of aerodynamic forces on oscillating three-dimensional wings at transonic speeds is the so-called kernel function method developed by the NASA Langley group consisting of Watkins, Runyan and Woolston (Refs. 70 and 59). This is a purely numerical method for solving the lifting-surface problem of unsteady subsonic and sonic flow. Although such a method leads to very complicated computations, it has no restrictions in planform so that for a complicated planform like one with swept trailing edges it may compare favorably with the exact solution described above.

The kernel function method is based on the following representation of the integral equation of lifting surface theory

$$w = \frac{1}{4\pi} \iint_S C_p(\xi, \eta, +0) \, \mathbf{K}(x - \xi, y - \eta) \, d\xi \, d\eta \tag{9.37}$$

The kernel function **K** is a complicated function of four variables $(x - \xi,$ $y - \eta,$ k and $M)$ and cannot be expressed in terms of known functions. It is singular at $x = \xi$ and $y = \eta$ and the singular part must therefore be split off and treated separately. For $M = 1$ the kernel is given by

$$\mathbf{K}(x, y) = k^2 e^{-i\bar{x}} \left\{ -\frac{1}{\bar{y}} K_1(\bar{y}) - \frac{\pi i}{2\bar{y}} \left[I_1(\bar{y}) - L_1(\bar{y}) - \frac{2}{\pi} \right] + \frac{1}{\bar{y}^2} \right.$$

$$\left. - \frac{2}{\bar{y}^2} \exp\left[\frac{i}{2} \left(\bar{x} - \frac{\bar{y}^2}{\bar{x}} \right) \right] + \frac{i}{\bar{y}^2} \int_{\bar{y}}^{\bar{x}} \exp\left[\frac{i}{2} \left(\lambda - \frac{\bar{y}^2}{\lambda} \right) \right] d\lambda \right\} \quad (9.38)$$

where $\bar{x} = kx$, $\bar{y} = ky$, K_1 the modified Bessel function in the conventional notation, and L_1 the modified Struve function of first order. For $x < 0$ the kernel is zero in accordance with the fact that there is no upstream influence in linearized transonic flow.

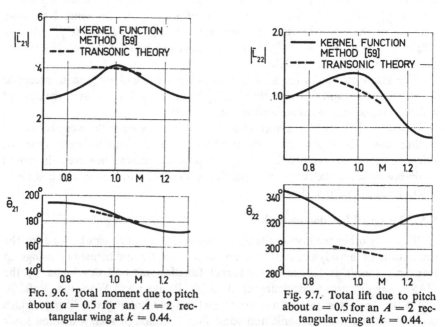

FIG. 9.6. Total moment due to pitch about $a = 0.5$ for an $A = 2$ rectangular wing at $k = 0.44$.

Fig. 9.7. Total lift due to pitch about $a = 0.5$ for an $A = 2$ rectangular wing at $k = 0.44$.

For an approximate solution of Eq. (9.37) C_p is first assumed to be expressible in a set of suitably chosen functions. For a rectangular wing the form selected in Ref. 59 was

$$C_p = 2\sqrt{(\sigma^2 - y^2)} \sum_{n=0}^{N} \sum_{m=0}^{M} a_{nm} x^{n-1/2} y^m \quad (9.39)$$

This loading is infinite at the leading edge, vanishes at $y = \pm\sigma$, and is finite at the trailing edge. For a wing with swept trailing edges one would need to select functions that are zero at the trailing edge. By substituting the assumed loading into Eq. (9.37) and carrying out the integration numerically w is expressed in terms of the unknown constants a_{nm}. By letting Eq. (9.37) be fulfilled in a suitable set of points on the wing a system of linear equations is obtained for the determination of the a_{nm}.

Examples of application of the kernel function method taken from Ref. 59 are given in Figs. 9.6 and 9.7. The coefficients calculated are the total lift and moment due to pitch about an axis at 50 per cent of the chord for an $A = 2$ rectangular wing oscillating with a reduced frequency of $k = 0.44$. The present notation for these coefficients is L_{21} and L_{22}, respectively. The subsonic and sonic results were calculated by use of the kernel function method, whereas the supersonic results were obtained from analytical methods. Also shown are results for the transonic region obtained from the exact theory of Chapter 7. As seen the agreement is quite good for the lift coefficient but not for the moment coefficient. The reason for this disagreement is not yet found.

CHAPTER 10

CONTROL SURFACE BUZZ

10.1. Introduction

PROBABLY the type of flutter that has most often been encountered by transonic airplanes to date is one-degree-of-freedom oscillation of a control surface, commonly known as "buzz". We recall that the linearized theory developed in previous chapters predicts negative hinge moment damping for control surfaces of high aspect ratios oscillating at low reduced frequencies, but this theory is incapable of providing an accurate description of the real phenomenon and can at most only be used for a qualitative analysis of the effects of certain parameters (such as the beneficial effect of a low control-surface aspect ratio demonstrated in Section 7.7). It is known that buzz is influenced by a number of factors like shock position, boundary layer separation, free play in the hinges, etc., in a manner not yet completely understood (see Ref. 55). In the early investigations of the phenomenon, as for example in that by Erickson and Stephenson (Ref. 15), it was found that buzz was always associated with oscillating shocks on the wing, located ahead of the hinge. However, it is now known† that, for a thin wing, buzz may also occur when the shocks are located on the control surface, as well as when they have moved back to the trailing edge.

In the latter case the position of the shocks will not oscillate and the flow over the control surface will be everywhere supersonic. A very simple approximate method of analysis, which has been proposed by Coupry and Piazzoli (Ref. 8), is then to assume a constant Mach number, equal to some suitably defined average Mach number, and apply the ordinary linearized supersonic theory. This theory predicts negative hinge moment damping in the two-dimensional case for Mach numbers less than $\sqrt{2}$ (with the hinge at the leading edge) so if the airfoil is thin, and hence the supersonic overvelocity over the control surface small, the simplified method predicts instability. We will not discuss this case any further, however, but will confine the rest of the chapter to the "classical" type of buzz as being the most important one from a practical point of view.

10.2. Unsteady transonic flow with shock waves

As pointed out in Chapter 1 the buzz problem cannot be analyzed successfully by use of the linearized transonic equation (1.36) for two reasons. Firstly, the reduced frequencies of interest based on control surface chord are usually so

† N. C. Lambourne, private communication.

110

low that the condition for linearization (1.40) is not fulfilled. Secondly, additional boundary conditions need be satisfied at the oscillating shock surfaces which were not considered in the discussion of the linearization process in Chapter 1.

Recently, two papers have appeared in which is treated the problem of unsteady transonic (non-viscous) flow with shock waves on the airfoil. In the work of Coupry and Piazzoli (Ref. 8) the velocity potential was used, whereas Eckhaus (Ref. 11) in his more recent paper works with the pressure or acceleration potential. The latter method has certain advantages since it permits also the entropy and vorticity fields to be calculated. We will here base our analysis on the velocity potential, since this in general gives somewhat simpler formulas, and hence follow the approach of Coupry and Piazzoli. These authors, however, were not able to give an analytic solution of the problem but suggested instead a numerical procedure for treating the subsonic region behind the shock. Eckhaus, on the other hand, was able to give a closed formula solution and we will therefore use his ideas to derive such a solution for the velocity potential as well. Also, with a slight extension of Eckhaus' method, we will give a result valid for all frequencies.

Assume that the unsteady flow is a small perturbation on the steady flow. Then, without limitations in frequency, the proper differential equation for the unsteady perturbation velocity potential is Eq. (1.33). With

$$\phi_2 = \text{Re}\,\{\varphi e^{ikt}\}$$

and restricting the analysis to two-dimensional flow it becomes

$$\frac{\partial}{\partial x}\{[1 - M^2 - M^2(\gamma + 1)\phi_{1x}]\varphi_x\} + \varphi_{zz} - 2ikM^2\varphi_x + k^2M^2\varphi = 0 \quad (10.1)$$

The last term can be neglected for slow oscillations. For rapid oscillations the field is governed by Eq. (1.36) as shown in Chapter 1. However, no great extra complication arises from retaining the last term in Eq. (10.1).

We will now consider the boundary conditions on an oscillating shock. For simplicity we assume the shock to be nearly straight and perpendicular to the free stream. Let the deviation of the shock position from its mean position x_s be given by

$$x - x_s = \Lambda(z, t) = \text{Re}\,\{\lambda(z)e^{ikt}\} \quad (10.2)$$

The first relation we make use of is that the tangential velocity component across the shock should be continuous. Hence

$$\phi_T{}^+ = \phi_T{}^- \quad (10.3)$$

where T denotes the tangential direction and $+$ and $-$ the values just downstream and upstream of the shock, respectively. Hence by integrating around the shock (which always must be finite since we assume the free stream velocity to be subsonic) we find that the velocity potential must be continuous across the instantaneous position of the shock. In accordance with the linearization of our problem we try to express this condition at the mean position, x_s, of the shock instead of at its instantaneous position. Expanding the total velocity potential just upstream of the shock in a Taylor series about $x = x_s$† and retaining only the lowest order terms in λ gives

$$\phi^- = \phi_1^-(x_s) + \lambda e^{ikt}\phi_{1x}^-(x_s) + \varphi^-(x_s)e^{ikt} \tag{10.4}$$

Expressing ϕ^+ in the same way and equating the values upstream and downstream of the shock we obtain for the jump of φ at $x = x_s$

$$\varphi^+ - \varphi^- = \lambda(\phi_{1x}^- - \phi_{1x}^+) \tag{10.5}$$

The second relation across the shock is found from the Rankine–Hugoniot conditions. A convenient form giving the velocity jump across the shock is the following one

$$\Delta U = \frac{2}{\gamma + 1} U\left(1 - \frac{c^2}{U^2}\right) \tag{10.6}$$

On the right-hand side either upstream or downstream values of c (velocity of sound) and U (velocity normal to the shock) may be used provided the sign of the jump is correctly chosen. Using Eq. (1.4) for c^2 and retaining only first-order terms in perturbation velocities gives

$$\phi_{1x}^- + \phi_{1x}^+ + e^{ikt}(\varphi_x^- + \varphi_x^+) = -\frac{2}{\gamma + 1}\left(1 - \frac{1}{M^2}\right)$$
$$+ \frac{4ik\lambda}{\gamma + 1}e^{ikt} - \frac{2(\gamma - 1)}{\gamma + 1}ik\varphi^- e^{ikt} \tag{10.7}$$

The last term comes from the ϕ_t-term in the equation for c^2, whereas the next last term arises from the motion of the shock. Eq. (10.7) is to be fulfilled at the instantaneous position of the shock. Transferring instead to the mean position $x = x_s$ by setting $\phi_{1x}^- + \lambda\phi_{1xx}^- \exp(ikt)$, etc., and equating terms proportional to $\exp(ikt)$ leads to‡

$$\lambda\left(\phi_{1xx}^- + \phi_{1xx}^+ - \frac{4ik}{\gamma + 1}\right) + \varphi_x^- + \varphi_x^+ = -\frac{2(\gamma - 1)}{\gamma + 1}ik\varphi^- \tag{10.8}$$

Equation (10.8) can be further simplified since it can be shown by use of the shock relation for the steady mean flow

$$\phi_{1x}^- + \phi_{1x}^+ = \frac{2}{\gamma + 1}\left(\frac{1}{M^2} - 1\right) \tag{10.9}$$

---

† This and the following requires analytic continuation of ϕ_1 through x_s on both sides of the shock as discussed by Coupry and Piazzoli (Ref. 8).
‡ The corresponding equation given by Coupry and Piazzoli is in error.

and the differential equation for ϕ_1 that, for a normal shock,

$$\phi_{1xx}^- + \phi_{1xx}^+ = 0 \qquad (10.10)$$

Hence, introducing this in Eq. (10.8) and substituting λ from Eq. (10.5) gives the following condition at $x = x_s$

$$\varphi_x^+ - i\alpha\varphi^+ = -\varphi_x^- - i\alpha\varphi^-[1 - \tfrac{1}{2}(\gamma - 1)(\phi_{1x}^- - \phi_{1x}^+)] \qquad (10.11)$$

where

$$\alpha = \frac{4k}{(\gamma + 1)(\phi_{1x}^- - \phi_{1x}^+)} \qquad (10.12)$$

The second term within the bracket on the right-hand side may be neglected since we assume small perturbations. Hence Eq. (10.11) may be simplified to

$$\varphi_x^+ - i\alpha\varphi_x^+ = -\varphi_x^- - i\alpha\varphi^- \qquad (10.13)$$

which is our final compatibility condition to be fulfilled at the mean position, $x = x_s$, of the shock. It is of interest to express α in terms of the Mach number, M_1, behind the shock. Using the thermodynamic relations and Eq. (10.9) we find that, to first order in $1 - M_1$,

$$\alpha = \frac{2kM^2}{1 - M_1^2} \simeq \frac{k}{1 - M_1} \qquad (10.14)$$

Hence, as $k/1 - M_1 \to \infty$, Eq. (10.13) tells that φ should be continuous across the shock and we arrive at the ordinary linearized problem as formulated in Chapter 1.

In addition to Eq. (10.13) the solution should also satisfy the ordinary linearized boundary conditions on the airfoil and in the wake. Thus the normal velocity is prescribed on the part of the x-axis occupied by the airfoil, and the pressure must be zero in the wake, as well as on the trailing edge (Kutta condition), i.e.

$$\varphi_z = w \qquad \text{on the airfoil} \qquad (10.15)$$

$$\varphi_x + ik\varphi = 0 \qquad \text{in the wake} \qquad (10.16)$$

where w is determined by the flap motion. For convenience we assume the leading edge and hinge of the flap to be located at $x = 0$ and the trailing edge at $x = 1$. Then, if only the flap oscillates,

$$w = \delta(1 + ikx) \qquad \text{for } 0 < x < 1 \qquad (10.17)$$

($\delta = $ flap amplitude) but zero on the rest of the airfoil. By requiring also that the solution should give velocities that vanish far from the airfoil in such a manner as to represent outgoing waves (radiation condition) we have completely specified our boundary value problem.

The only approximation to the exact linearized problem made so far is the assumption of a normal shock. To simplify the problem further we make the additional assumption that $\phi_{1x} = $ constant in the region downstream of the shock. Setting the Mach number equal to M_1 in the region, φ should then satisfy the acoustical equation for subsonic flow, i.e.

$$(1 - M_1{}^2)\varphi_{xx} + \varphi_{zz} - 2ikM_1{}^2\varphi_x + k^2 M_1{}^2\varphi = 0 \qquad (10.18)$$

The coefficients in the last two terms differ somewhat from those of Eq. (10.1) but since both M and M_1 are $1 + 0(\phi_{1x})$ the difference is unimportant, as follows from Chapter 1.

In addition, we will assume the airfoil to be at zero angle of attack (so that φ is antisymmetric in z) and the shock to be located ahead of the hinge (i.e. $x_s < 0$). Thus, the right-hand side of Eq. (10.13) is zero and the boundary conditions for the problem become

$$\varphi_x(x_s, z) - i\alpha\varphi(x_s, z) = 0 \qquad (10.19)$$

$$\varphi_z(x, 0) = w \qquad \text{for } x < 1 \qquad (10.20)$$

$$\varphi_x(x, +0) + ik\varphi(x, +0) = 0 \qquad \text{for } x \geq 1 \qquad (10.21)$$

Following Eckhaus' (Ref. 11) idea, we will seek a solution to the simplified boundary value problem defined by Eqs. (10.18)–(10.21) by considering the modified potential function $\bar{\varphi}$ defined by

$$\bar{\varphi} = \varphi_x - i\alpha\varphi \qquad (10.22)$$

With $\bar{\varphi}$ known, φ can then be found from the formula

$$\varphi = -\int_x^\infty \bar{\varphi}(u, z)e^{-i\alpha(u-x)}\, du \qquad (10.23)$$

(Here we have assumed that $\varphi(\infty, z) = 0$ which requires that $Im(k) < 0$.)

Now let us for the moment disregard the boundary condition in the wake, Eq. (10.21), and seek a solution fulfilling the remaining boundary conditions. The boundary value problem for $\bar{\varphi}$ thus becomes

$$(1 - M_1{}^2)\bar{\varphi}_{xx} + \bar{\varphi}_{zz} - 2ikM_1{}^2\bar{\varphi}_x + k^2 M_1{}^2\bar{\varphi} = 0 \qquad (10.24)$$

$$\bar{\varphi}(x_s, z) = 0 \qquad (10.25)$$

$$\bar{\varphi}_z(x, 0) = w_x - i\alpha w = \bar{w} \qquad (10.26)$$

We may obtain the required solution by constructing one which is antisymmetric about $x = x_s$ and has a given normal derivative on $z = 0$. First we set

$$\bar{\varphi} = \bar{\varphi}_1 e^{i\kappa M_1 x} \qquad (10.27)$$

where

$$\kappa = \frac{kM_1}{1 - M_1{}^2} = \frac{kM_1}{\beta^2} \qquad (10.28)$$

Then $\bar\varphi_1$ will be governed by the wave equation, i.e.

$$\bar\varphi_{1xx} + \frac{1}{\beta^2}\bar\varphi_{1zz} + \kappa^2\bar\varphi_1 = 0 \qquad (10.29)$$

which is symmetric in x so that a solution to this may be obtained from another solution by replacing x with $-x$. The boundary condition (10.25) will remain unchanged whereas that given by Eq. (10.26) will be replaced by

$$\bar\varphi_{1z}(x, 0) = \bar{w} \exp(-i\kappa M_1 x) = \bar{w}_1 \qquad (10.30)$$

The fundamental solution with a given normal derivative on $z = 0$ for $x > 0$ is

$$\bar\varphi_1 = \int_0^\infty G(x - \xi, z)\bar{w}_1(\xi)\,d\xi \qquad (10.31)$$

where

$$G(x, z) = \frac{i}{2\beta} H_0^{(2)}[\kappa\sqrt{(x^2 + \beta^2 z^2)}] \qquad (10.32)$$

To make $\bar\varphi_1$ zero for $x = x_s$ we subtract the mirror image with respect to $x = x_s$. This is achieved by replacing $G(x - \xi, z)$ in Eq. (10.31) by $K(x, \xi, z)$, where

$$K(x, \xi, z) = G(x - \xi, z) - G(2x_s - x - \xi, z) \qquad (10.33)$$

Hence we obtain after introducing Eq. (10.27):

$$\bar\varphi = \int_0^\infty K(x, \xi, z) \exp[i\kappa M_1(x - \xi)]\bar{w}(\xi)\,d\xi \qquad (10.34)$$

From Eq. (10.23) we may now calculate φ. After two integrations by parts and interchanging the order of integration the result may be cast in the following form:

$$\varphi = \int_0^\infty \exp[i\kappa M_1(x - \xi)]K^* w\,d\xi \qquad (10.35)$$

where

$$K^*(x, \xi, z) = G(x - \xi, z) + G(2x_s - x - \xi, z)$$

$$-2i(\alpha - \kappa M_1)\int_x^\infty \exp[-i(u - x)(\alpha - \kappa M_1)]G(2x_s - u - \xi, z)\,du$$

$$\equiv G(x - \xi, z) + \bar{K}^*(x + \xi - 2x_s, z) \qquad (10.36)$$

This solution satisfies the differential equation and all boundary conditions except that in the wake, Eq. (10.21). From Eq. (10.35) and (10.21) an integral equation can be written down to solve the unknown w for $x > 1$. In Eckhaus' (Ref. 11) treatment of the problem he showed how this equation could be solved approximately for large values of κ. However, we will instead employ the previously used method of successive cancellations (Schwartzschild's method) to make the wake pressure vanish, since this method has no limitation in frequency, except that the frequency must not be zero. Thus, any suitable distribution of w for $x > 1$ may first be chosen in Eq. (10.35). This gives a solution $\varphi^{(0)}$ with a pressure coefficient

$$C_p^{(0)} = -2(\varphi_x^{(0)} + ik\varphi^{(0)})e^{ikt} \qquad (10.37)$$

that in general is not zero in the wake. Hence we need to add a solution which in the wake takes the value $-C_p^{(0)}$ but does not change the remaining boundary conditions. Now we note that $C_p(x, z, t)$, as well, is a solution of Eq. (10.18) so, with the aid of Schwartzschild's kernel (see Section 7.4), we may write a solution with the required value in the wake as follows

$$C_p^{(1)} = -\int_1^\infty \exp\left[i\kappa M_1(x - \xi)\right]G_1(x, \xi, z)C_p^{(0)}(\xi)\, d\xi \qquad (10.38)$$

For $z = 0$ and $x < 1$, G_1 is given by the simple expression

$$G_1 = \frac{1}{\pi}\sqrt{\left(\frac{1 - x}{\xi - 1}\right)}\frac{e^{-i\kappa(\xi - x)}}{\xi - x} \qquad (10.39)$$

Furthermore, for $x < 1$,

$$G_{1z}(x, \xi, 0) = 0 \qquad (10.40)$$

It is easily shown that $C_p^{(1)}$ satisfies the Kutta condition. Now we investigate whether the remaining boundary conditions are unchanged. For this we calculate the velocity potential corresponding to Eq. (10.38) by use of the formula

$$\varphi = -\tfrac{1}{2}e^{-ikt}\int_{-\infty}^x C_p(v, z, t + v - x)\, dv \qquad (10.41)$$

Upon changing the order of integration we obtain

$$\varphi^{(1)} = \tfrac{1}{2}e^{-ikt}\int_1^\infty \exp\left[i\kappa M_1(x - \xi)\right]\bar{G}_1 C_p^{(0)}\, d\xi \qquad (10.42)$$

where

$$\bar{G}_1(x, \xi, z) = \int_{-\infty}^x \exp\left[-i(k + \kappa M_1)(x - v)\right]G_1(v, \xi, z)\, dv \qquad (10.43)$$

In view of Eq. (10.40) we see that $\varphi_z^{(1)}(x, 0) = 0$ for $x < 1$ as required. However, the solution does not satisfy the compatibility condition at the shock, Eq. (10.19). In analogy with Eq. (10.35) we therefore try the solution

$$\varphi^{(1)} = \tfrac{1}{2}e^{-ikt}\int_1^\infty \exp\left[i\kappa M_1(x - \xi)\right]K_1(x, \xi, z)C_p^{(0)}(\xi)\, d\xi \qquad (10.44)$$

where

$$K_1 = \bar{G}_1(x, \xi, z) + \bar{G}(2x_s - x, \xi, z)$$

$$-2i(\alpha - \kappa M_1) \int_x^\infty \exp\left[-i(u - x)(\alpha - \kappa M_1)\right]\bar{G}_1(2x_s - u, \xi, z)\, du \quad (10.45)$$

This solution can be easily shown to satisfy Eq. (10.19) but does not give the correct pressure in the wake. After some manipulations we may cast the result for $C_p^{(1)}$ in the following form:

$$C_p^{(1)} = -\int_1^\infty \exp\left[i\kappa M_1(x - \xi)\right]K_1^*(x, \xi, z)C_p^{(0)}(\xi)\, d\xi \quad (10.46)$$

where

$$K_1^* = G_1(x, \xi, z) + \bar{R}_1^*(2x_s - x_1\, \xi, z) \quad (10.47)$$

and

$$\bar{R}_1^*(x, \xi, z) = -G_1(x, \xi, z) + 2i(\alpha + k)\int_0^\infty \left\{\left(1 - \frac{\alpha - \kappa M_1}{\alpha - 2\kappa M_1 - k}\right)\right.$$

$$\times \exp\left[-i(k + \kappa M_1)v\right] + \frac{\alpha - \kappa M_1}{\alpha - 2\kappa M_1 - k}$$

$$\left.\times \exp\left[-i(\alpha - \kappa M_1)v\right]\right\}G_1(x - v, \xi, z)\, dv \quad (10.48)$$

In the wake the total pressure will hence be

$$C_p = C_p^{(0)} + C_p^{(1)} = \bar{C}_p^{(1)} \quad (10.49)$$

where

$$\bar{C}_p^{(1)} = -\int_1^\infty \exp\left[i\kappa M_1(x - \xi)\right]\bar{R}_1^*(2x_s - x, \xi, 0)C_p^{(0)}(\xi)\, d\xi \quad (10.50)$$

This vanishes only when the shock recedes upstream to infinity, so we reduce the pressure to zero by successive application of Eq. (10.46). This leads, finally, to the following series for C_p

$$C_p = \sum_0^\infty C_p^{(0)} \quad (10.51)$$

where $C_p^{(0)}$ and $C_p^{(1)}$ are given by Eqs. (10.37) and (10.46), respectively, and where

$$C_p^{(n)} = -\int_1^\infty \exp\left[i\kappa M_1(x - \xi)\right]K_1^*(x, \xi, 0)\bar{C}_p^{(n-1)}(\xi)\, d\xi \quad (10.52)$$

$$\bar{C}_p^{(n)} = -\int_1^\infty \exp\left[i\kappa M_1(x - \xi)\right]\bar{R}_1^*(x, \xi, 0)\bar{C}_p^{(n-1)}(\xi)\, d\xi \quad (10.53)$$

with $\bar{C}_p^{(1)}$ given by Eq. (10.50) above. The series converges for all non-zero values of κ and the distance between the shock and the trailing edge. It is likely that the convergence in most practical cases is very rapid so that only the first two terms need be retained.

The effect of the shock is contained in the kernel parts K^* and K_1^* which will now be further discussed. We start with K_1^*. Integrating Eq. (10.48) once by parts gives

$$\pi\sqrt{(\xi-1)}\, e^{i\kappa(\xi-x)}K_1^*(x,\xi,0) = A\frac{\sqrt{(1-x)}}{\xi-x} - \frac{2(\alpha+k)}{a}\int_0^\infty e^{-iva}\, dv$$

$$\times\left[1 - \frac{\alpha-\kappa M_1}{b} + \frac{a\,(\alpha-\kappa M_1)}{b\,(a+b)}e^{-ibv}\right]\frac{\partial}{\partial v}\left(\frac{\sqrt{(1-x+v)}}{\xi-x+v}\right) \quad (10.54)$$

where

$$A = \frac{2(\alpha+k)\kappa}{a(a+b)} - 1$$

$$a = k + \kappa(1+M_1) \quad (10.55)$$

$$b = \alpha - k - 2\kappa M_1$$

Now we should neglect terms of order $1 - M_1$ and higher. Hence, from Eqs. (10.14) and (10.28) it follows that

$$\kappa = \tfrac{1}{2}\alpha + 0(1-M_1) \quad (10.56)$$

which is introduced into the expression for A. Then we see that

$$A = 0(1-M_1)$$

which hence could be neglected. Continuing in this manner, each time neglecting all higher-order terms in the out-integrated parts, we obtain after two more integrations by parts

$$\pi\sqrt{(1-\xi)}e^{i\kappa(\xi-x)}K_1^*(x,\xi,0) = \frac{1}{\alpha^2}\frac{\partial^2}{\partial x^2}\left(\frac{\sqrt{(1-x)}}{\xi-x}\right) - \frac{2(\alpha+k)}{a^3}\int_0^\infty e^{-iav}\, dv$$

$$\times\left[1 - \frac{\alpha-\kappa M_1}{b} + \frac{a^3\,(\alpha-\kappa M_1)}{b\,(a+b)^3}e^{-ibv}\right]\frac{\partial^3}{\partial v^3}\left[\frac{\sqrt{(1-x+v)}}{\xi-x+v}\right] \quad (10.57)$$

From Eqs. (10.55) and (10.56) it follows that $b = 0(k)$ so that for small k we may in the square bracket set

$$e^{-ibv} \simeq 1 - ibv$$

On the other hand, if k is large, of order unity or more, the whole integral will be of order $(1-M_1)^2$ so that this approximation is good for all k. Then, finally, neglecting higher order terms in $1 - M_1$

$$\pi\sqrt{(1-\xi)}e^{-i\kappa(\xi-x)}K_1^*(x,\xi,0) = \frac{1}{\alpha^2}\frac{\partial^2}{\partial x^2}\left(\frac{\sqrt{(1-x)}}{\xi-x}\right)$$

$$+ \frac{1}{\alpha^2}\int_0^\infty e^{-iav}(1+iav)\frac{\partial^3}{\partial v^3}\left(\frac{\sqrt{(1-x+v)}}{\xi-x+v}\right)dv \quad (10.58)$$

In a similar manner the expression for $\bar{R}^*(x, 0)$ may be simplified to

$$e^{i\kappa x}\bar{R}^*(x, 0) = -\frac{i}{2\beta}\int_0^\infty e^{-i\alpha v}\frac{\partial}{\partial v}\left\{e^{i\kappa(v+x)}H_0^{(2)}\left[\kappa(v + x)\right]\right\}dv \quad (10.59)$$

Eqs. (10.58) and (10.59) are suitable for investigating the behavior of $\bar{R}^*$ and $\bar{R}_1^*$ as $M_1 \to 1$ for a fixed k, i.e. as $\alpha = k/(1 - M_1) \to \infty$. Since these kernel parts are non-singular ($x + \xi - 2x_s$ is always positive) we may easily obtain any number of terms in an asymptotic expansion through integration by parts. In this manner we find that

$$\bar{R}^*/K^* = 0(\alpha^{-1}) \quad (10.60)$$

and

$$\bar{R}_1^*/K_1^* = 0(\alpha^{-2}) \quad (10.61)$$

Hence we see that the reflection at the shock, which is represented by $\bar{R}^*$ and $\bar{R}_1^*$, gives at most terms of relative order α^{-1}. On the other hand, the upstream influence from the wake goes as $\alpha^{-1/2}$ (see Section 2.4); consequently the proper first-order problem (in which terms of α^{-1} and higher are neglected but terms of order $\alpha^{-1/2}$ are retained) is that for which $\bar{R}^*$ and $\bar{R}_1^*$ are neglected, i.e. the problem for an airfoil without a shock but which is infinite upstream. Hence the first-order effect of the shock is simply to make the airfoil look semi-infinite. This important conclusion reached by Eckhaus does not hold if the shock is too close to the hinge since then the asymptotic expansions for $\bar{R}^*$ and $\bar{R}_1^*$ break down.

That the shock acts as an effective absorber of unsteady disturbances coming from downstream can be demonstrated much more simply by considering the one-dimensional case. The solution for a receding one-dimensional wave with a unit amplitude reads (see Chapter 1)

$$\exp\left[\frac{ikM_1x}{1 - M_1}\right] \quad (10.62)$$

The reflection at the shock will cause an advancing wave with the solution

$$B\exp\left[-\frac{ikM_1x}{1 + M_1}\right] \quad (10.63)$$

The constant B is found by postulating that the sum of (10.62) and (10.63) should satisfy the shock relation (10.13). This gives

$$B = \frac{kM_1 - \alpha(1 - M_1)}{kM_1 + \alpha(1 + M_1)} = 0(1 - M_1) \quad (10.64)$$

Hence the strength of reflected signal will only be of order $1 - M_1 = 0(\phi_{1x})$ of the incoming signal. (Since terms of this order were neglected in deriving our basic equations it is not possible by use of potential theory to find the exact value of B.)

The above indicates that, in the case of large α, and the shock not too close to the hinge, the presence of the shock can probably not cause phase lags sufficiently large to lead to negative damping of the control surface. For α small, and when the shock is close to the hinge, further calculations must be awaited until definite conclusions can be reached.†

Extension of the theory for the case of the shock located on the control surface is possible in the manner shown by Eckhaus (Ref. 11).

A two-dimensional theory like the one presented will of course only be of limited practical value since, as shown in previous chapters, three-dimensional effects are always very large at transonic speeds. It is possible to extend the theory to an inboard rectangular control surface by use of Fourier transformation in the span-wise direction, but since the errors due to the simplifications introduced are unknown (in particular, the best choice of M_1 is an open question) such an effort seems hardly worth-while at present.

10.3. Influence of the boundary layer

Since the preceding section indicated that reflections of disturbances by the shock are weak, we are led to look for other possible mechanisms whereby negative control-surface damping can arise. So far we have not considered the effect of the boundary layer in our discussion. It is well known that a shock wave of sufficient strength impinging on a boundary layer will induce separation and that such shock-wave–boundary-layer interaction can alter considerably the steady-state transonic pressure distribution around a wing or a body.

Already in the early investigations of buzz it was found that separation was involved, but the precise role that this was playing was not known. In a recent report by Lambourne (Ref. 29) the close correlation between the occurrence of separation and the onset of buzz is clearly demonstrated. His experiments show that the critical Mach number for buzz coincides very closely with that for which the separated region has spread markedly over the control surface. On basis of his experiments, Lambourne also suggests a possible way in which the separation can cause negative damping. We will here briefly review his findings and the underlying experimental evidence.

In attempting an explanation of the phenomenon it is expedient to consider firstly the effects of unsteady flow on the separation, secondly the effect of the changes in the separation on the aerodynamic forces acting on the control surface.

For simplicity let us consider an airfoil–control-surface combination at a small angle of attack with shock-induced separation on the upper surface only,

† Note added in proof: Recent, as yet unpublished calculations by Eckhaus show, however, that there is a region of negative image moment damping for $M_1 > 0.95$.

and the shock located well ahead of the hinge. Now assume that the control surface starts impulsively an upward rotation from its undeflected position. This motion will create a positive pressure wave, in the first instance in phase with, and of a strength proportional to, the angular velocity of the control surface. After a certain time lag, which depends on the difference between local sound speed and fluid velocity, the pressure wave will reach the shock. The shock will then adjust its speed so as to correspond to the increased pressure ratio, i.e. it will move upstream with a velocity proportional to the strength of the pressure wave. The effect of the control surface motion will thus be to increase the shock strength, and the increased shock strength will in turn increase the severity of the separation.

Similarly, a deflection of the control surface downwards will cause a negative pressure wave that will weaken the shock strength and make the separation less severe. This description of the effect of control surface motion on the shock-induced separation is supported in the main by Lambourne's measurements of shock position and thickness of the separated layer as functions of time. These show that the separation is most severe just after the control surface attains its maximum upward velocity, whereas during most of the downstroke the flow is attached. Also, the shock position and control surface deflection are essentially in phase, i.e. the shock velocity and the angular velocity of the control surface are also in phase.

Now consider the effect of the change in the separation on the forces and moments acting on the control surface. Steady-state measurements indicate that the onset of separation on either surface of the airfoil gives an incremental hinge moment that tends to turn the control surface towards the separation. Thus, separation on the upper surface tends to turn the control surface upwards. Assuming that this remains valid also for unsteady flow, it follows that the change in hinge moment due to change in the separation will almost always act in the direction of the control surface motion so that energy may be absorbed from the free stream and cause unstable oscillations.

It will be noted that the essential point in the above description of the mechanism is the alleged effect of unsteady flow on the shock-induced separation. A closer study of this particular phenomenon is certainly required to verify the explanation given and to obtain a deeper understanding of the problem. In this connection we should mention a paper by Trilling (Ref. 67) in which is treated the problem of the interaction of an oblique shock with a laminar boundary layer. He finds that under certain conditions self-maintained oscillations may occur. Possibly, such a mechanism could be involved in transonic buffeting. A similar analysis for a normal shock interacting with a turbulent boundary layer could possibly give important quantitative information regarding the response of the separation to pressure disturbances coming from the region downstream of the shock. Finally, it should be mentioned that the shock itself is not essential to the phenomenon, since Lambourne shows that buzz may also occur for an airfoil at high angle of attack with ordinary leading edge separation, but at insufficient speed for shock waves to occur. Thus, the essential feature of the buzz instability is probably the separation and not the shock wave, and the discussed change in separation through the varying shock strength seems to be only a particular example of the general phenomenon. Further studies of the effect of unsteady flow on the separation will possibly lead to a better understanding of this and related problems.

CHAPTER 11

EXPERIMENTAL DETERMINATION OF AIR FORCES ON OSCILLATING WINGS AT TRANSONIC SPEEDS

11.1. Introduction

IN THIS chapter we have collected available experimental results for such cases where direct comparisons with theory are possible. The unclassified literature on experimentally determined unsteady air forces at transonic speeds is not very extensive, however, no doubt partly due to the special difficulties associated with transonic testing. A particular difficulty in connection with high speed testing of oscillating wings in general is the very high frequency necessary to obtain a moderately high reduced frequency. This is especially unfortunate at transonic speeds, since successful comparisons with the linearized transonic theory require a high reduced frequency. Then, of course, wind tunnel wall interference effects also pose severe problems at transonic speeds which will be discussed subsequently.

11.2. Wind tunnel wall interference

Wind tunnel wall interference for an oscillating airfoil has been computed by Woolston and Runyan (Ref. 71) for subsonic speeds, and by Miles (Ref. 48) for supersonic speeds. Extensions to porous walls were given by Drake (Ref. 9). The important result obtained from the subsonic analysis is the occurrence of resonance when disturbances require an odd integer number of oscillation half-periods to travel from the model to the wall and back. If we consider the airfoil to be situated midway between the walls the resonance frequency† is given by

$$\frac{\omega H}{c} = (2n + 1)\pi\sqrt{(1 - M^2)} \tag{11.1}$$

where H is the test section height.

At supersonic speeds resonance does not occur but interference effects may nevertheless be quite large (Refs. 9, 48).

Naturally, these linearized analyses do not apply at transonic speeds. However, from Eq. (11.1) it seems possible that resonance may occur at high subsonic Mach numbers even for low frequencies. Now a slotted wind tunnel wall might alleviate this interference effect (Drake's paper on porous walls does not indicate whether resonance is possible) but it is not unlikely that resonance

† According to Miles (Ref. 48) a more appropriate term would be "cut-off frequency".

can occur, even if steady-state wall interference is minimized by the slotted walls. In the low supersonic speed region the problem of shock reflection from the walls is still not satisfactorily solved in the steady-flow case.

The simplest method to estimate the wall interference effects is to test the same model in two or more wind tunnels of different test section height. This was done by Orlick-Rückemann and Olsson in Ref. 53. They found that the interference effects were noticeable but in general small.

11.3. Experimental methods

Of the very few experimental investigations of air forces on oscillating wings at transonic speeds that have been published to date practically all have concerned measurements of the damping in pitch. Two different experimental techniques have been used, namely the free and forced oscillation techniques, respectively. In the free oscillation technique the model is elastically mounted so that it can perform pitching oscillations around a fixed axis. After excitation the logarithmic decrement of the decaying oscillation is measured. Orlick-Rückemann and Olsson used an apparatus, called the "Dampometer", which makes the evaluation of the logarithmic decrement automatic. In the forced-oscillation technique the power required to maintain an oscillation with constant amplitude is measured. The model is oscillated at the resonance frequency of the set-up (with wind on). The forced-oscillation technique requires a fairly complicated mechanism for the excitation of the model but the method is advantageous at transonic speeds since the non-linear effect of amplitude can be investigated.

11.4. Results and comparisons with theory

In Fig. 11.1 are shown results obtained by Orlik-Rückemann and Olsson (Ref. 53) for the damping-in-pitch at transonic and supersonic speeds for a triangular wing of $A = 1.45$ oscillating at the frequencies 70 c/s and 94 c/s. This corresponds at $M = 1$ to $k = 0.30$ and 0.37, respectively. The pitching axis is located at 60 per cent of the root chord. Also shown in the figure are the theoretical result at $M = 1$ from the low aspect ratio theory of Chapter 3, Eq. (3.38), and results according to the supersonic theory (Ref. 69). The indicated Mach number variation around $M = 1$ is obtained from the similarity law given in Section 1.8. When the higher order frequency terms are retained in the transonic theory the results for $k = 0.30$ and 0.37 become indistinguishable. As seen, the agreement is quite good in the lower transonic region, but the theory fails to reproduce the rapid drop in damping that occurs for M slightly above unity. The large discrepancy between theory and experiments at supersonic speeds is difficult to explain. A possible explanation could be that it is due to the half-model technique used. This explanation is supported by the investigation of Tobak (Ref. 66) who measured the damping on full models of

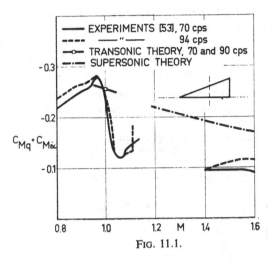

FIG. 11.1.

low aspect ratio wing–body combinations at subsonic and supersonic speeds. Some of his results are incorporated in Fig. 11.2. The theory shown for supersonic Mach numbers is the "modified theory" of Ref. 72, i.e. where the influence of the body is calculated from a low aspect ratio theory corresponding to that

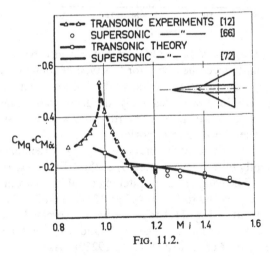

FIG. 11.2.

given in Chapter 5. As seen, the agreement is quite good. At transonic speed, for which measurements on the same models were made by Emerson and Robinson (Ref. 12), the agreement with the theory of Chapters 3 and 5 is fairly good, except that the theory does not show the very sharp peak in the damping at $M = 0.98$. This peak is most likely caused by non-linear effects since the reduced frequency, estimated to $k = 0.18$ at $M = 1$, is probably not sufficiently high for the linearized theory to be valid. However, one should also notice that tunnel resonance should occur at $M = 0.98$ according to Eq. (11.1).

In Ref. 12 were also given results for a wing–body combination with an unswept tapered wing of $A = 3$. No theoretical results for this configuration have been computed. However, the very rapid and erratic variation of the measured damping with M in the transonic region, shows that non-linear effects must be large in this case, and the linearized theory is therefore likely to give

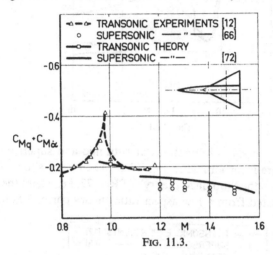

FIG. 11.3.

poor results. From the reasoning in Chapter 1 it can also be concluded that non-linear effects should indeed be larger for this type of configuration than for the $A = 2$ delta-wing configuration. Firstly, the reduced frequency is lower (about 0.1) since the chord is lower. Secondly, the aspect ratio is larger making the non-lifting flow disturbances larger. Thirdly, the wing is unswept so that transonic effects in general are more important.

Finally, we will show some preliminary results for a rectangular wing of $A = 2$ obtained recently by J. B. Bratt (Ref. 5) of the NPL by use of the forced-oscillation technique. The wing performs pitching oscillations around $x_0 = 0.42$ with a frequency corresponding to $k = 0.445$ at $M = 1$ and 0.508 at $M = 0.85$. The results are presented in Fig. 11.4 together with theoretical ones according to the transonic theory described in Chapter 7, and also some results from a recent subsonic theory by Acum (British ARC 19229). As seen, the agreement with the transonic theory is poor, in particular if compared to the subsonic one. Apparently the transonic theory in this case requires a higher reduced frequency, or a thinner wing (the wing used is 10 per cent thick) to be valid.

In summary it follows from the few comparisons shown that the linearized transonic theory gives reasonably good agreement with experiments for low aspect ratio delta wings, even at low reduced frequencies, whereas the scant informations for rectangular wings indicates poor agreement, even at moderately large frequencies. That the linearized theory works better for pointed wings than for straight wings is to be expected since, as discussed in Chapter 1, the

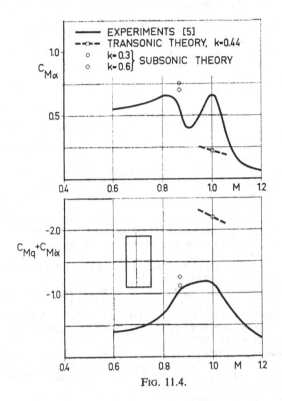

FIG. 11.4.

receding-wave part is much stronger for a straight wing. A more complete exploration of the limits of applicability of the linearized theory requires large systematic investigations in which frequency and wing shape are varied.

REFERENCES

1. ADAMS, M. C. and SEARS, W. R., Slender-Body Theory—Review and Extension. *J. Aero. Sci.*, **20**, No. 2, 85–98, 1953.
2. ANONYMOUS, Tables of Aerodynamic Coefficients for an Oscillating Wing-Flap System in a Subsonic Compressible Flow, NLL Report F.151, 1954.
3. BATEMAN, H., *Partial Differential Equations of Mathematical Physics*, New York, Dover Publications, 1944.
4. BEHRBOHM, H., Auftriebslöschung und Aufwindlöschung, zwei duale Methoden der linearisierten Überschalltragflächentheorie. Anwendung auf den Doppeldreicksflügel. *Z. Flugwiss.*, **4**, No. 8, 263–268, 1956.
5. BRATT, J. B. Unpublished experiments.

6. BURGER, A. P., On the Asymptotic Solution of Wave Propagation and Oscillation Problems, NLL Report F.157, 1954.
7. COLE, J. D. and MESSITER, A. F., Expansion Procedures and Similarity Laws for Transonic Flow—I. Slender Bodies at Zero Incidence, OSR TN No. 56-1, 1956.
8. COUPRY, G. and PIAZZOLI, G., Étude du Flottement en Regime Transonique. *Rech. Aéro.*, No. 63, 1958.
9. DRAKE, D. G., The Motion of an Oscillating Airfoil in a Compressible Free Jet. *J. R. Aero. Soc.*, **60**, No. 9, 621–623, 1956.
10. ECKHAUS, W., Asymptotic Solution of the Two-Dimensional Oscillating Airfoil Problem, for High Subsonic Mach Numbers. Paper presented at *Ninth International Congress of Applied Mechanics, Brussels*, 1956.
11. ECKHAUS, W., Two-Dimensional Transonic Unsteady Flow with Shock-Waves, OSR TR No. 59-591, 1959.
12. EMERSON, H. F. and ROBINSON, R. C., Experimental Wind-Tunnel Investigation of the Transonic Damping-In-Pitch Characteristics of Two Wing-Body Combinations, NASA Memo. 11-30-58A, 1958.
13. ERDÉLYI, A. *et al.*, *Higher Transcendental Functions* Vol. I. Bateman Manuscript Project, McGraw-Hill, New York, 1953.
14. ERDÉLYI, A., *Asymptotic Expansions*, New York, Dover Publications, 1956.
15. ERICKSON, A. L. and STEPHENSON, J. D., A Suggested Method of Analyzing for Transonic Flutter of Control Surfaces based on Available Experimental Evidence, NACA RM No. A7F30, 1947.
16. GARRICK, I. E. and RUBINOW, S., Flutter and Oscillating Air-Force Calculations for an Airfoil in a Two-Dimensional Supersonic Flow, NACA Report 846, 1946.
17. GARRICK, I. E., Some Research on High-Speed Flutter. *Third Anglo-American Aeronautical Conference, Brighton*, 1951.
18. GARRICK, I. E., Some Concepts and Problem Areas in Aircraft Flutter, The 1957 Minta Martin Aeronautical Lecture, IAS SMF Fund Paper No. FF-15.
19. GERMAIN, P. and BADER, R., Quelques Remarques sur les Mouvements Vibratoires d'une Aile en Régime Supersonique. *Rech. Aéro.*, No. 11, 3–13, 1949.
20. GUNN, J. C., Linearized Supersonic Aerofoil Theory. *Philos. Trans.*, A **240**, 327–373, 1947.
21. HASKIND, M. D. and FALKOVICH, S. V., Vibration of a Wing of Finite Span in a Supersonic Flow. *Prikl. Mat. i Mekh.*, **11**, 1947. (Translated in NACA TM No. 1257, 1950.)
22. HJELTE, F., Methods for Calculating Pressure Distributions on Oscillating Wings of Delta Type at Supersonic and Transonic Speeds, Stockholm, KTH AERO TN 39, 1956.
23. JONES, R. T., Properties of Low Aspect Ratio Pointed Wings at Speeds Below and Above the Speed of Sound, NACA Report No. 835, 1946.
24. JORDAN, P. F., Aerodynamic Flutter Coefficients for Subsonic, Sonic and Supersonic Flow (Linear Two-Dimensional Theory), ARC R & M No. 2932, 1957.
25. KANTROWITZ, A., The Formation and Stability of Normal Shock Waves in Channel Flows, NACA TN 1225, 1947.
26. KENNET, H., ASHLEY, H. and STAPLEFORD, R. L., Forces and Moments on Oscillating Slender Wing-Body Combinations at Supersonic Speed—II. Applications and Comparison with Experiments, OSR TN No. 58-114, 1958.
27. KUO, Y. H., On the Stability of Two-Dimensional Smooth Transonic Flows. *J. Aero. Sci.*, **18**, No. 1, 1–6, 1951.
28. LAMB, H., On Sommerfeld's Diffraction Problem; and on Reflection by a Parabolic Mirror. *Proc. Lond. Math. Soc.*, Ser. 2, **4**, 190–203, 1907.
29. LAMBOURNE, N. C., Some Instabilities Arising From the Interactions Between Shock Waves and Boundary Layers, National Physical Laboratory NPL/Aero/348, 1958.
30. LANDAHL, M. T., The Flow Around Oscillating Low Aspect Ratio Wings at Transonic Speeds, Stockholm, KTH AERO TN 40, 1954.
31. LANDAHL, M. T., Forces and Moments on Oscillating Slender Wing-Body Combinations at Sonic Speed, OSR TN No. 56-109, 1956.
32. LANDAHL, M. T., A Strip Theory for Oscillating Thin Wings with Subsonic Leading and Supersonic Trailing Edges, OSR TN No. 56-155, 1956.

33. LANDAHL, M. T., Forces and Moments on Oscillating Low Aspect Ratio Wings and Wing–Body Combinations at Transonic Speeds. Paper presented at *Ninth International Congress of Applied Mechanics, Brussels*, 1956.
34. LANDAHL, M. T., Aerodynamic Derivatives for Oscillating Three-Dimensional Wings in Transonic Flow. Paper presented at *First International Congress on the Aeronautical Sciences, Madrid*, 1958.
35. LANDAHL, M. T., Theoretical Studies of Unsteady Transonic Flow—I. Linearization of Equations of Motion, Aeronautical Research Institute of Sweden (FFA), Report 77, 1958.
36. LANDAHL, M. T., Theoretical Studies of Unsteady Transonic Flow—II. The Oscillating Semi-Infinite Rectangular Wing, Aeronautical Research Institute of Sweden (FFA), Report 78, 1958.
37. LANDAHL, M. T., Theoretical Studies of Unsteady Transonic Flow—III. The Oscillating Low Aspect Ratio Rectangular Wing, Aeronautical Research Institute of Sweden (FFA), Report 79, 1958.
38. LANDAHL, M. T., Theoretical Studies of Unsteady Transonic Flow—IV. The Oscillating Rectangular Wing with Control Surface, Aeronautical Research Institute of Sweden (FFA), Report 80, 1958.
39. LANDAHL, M. T., Theoretical Studies of Unsteady Transonic Flow—V. Solution for the Delta Wing and Wings of General Polygonal Planforms, Aeronautical Research Institute of Sweden (FFA), Report 81, 1959.
40. LANDAHL, M. T., ASHLEY, H. and MOLLÖ-CHRISTENSEN, E., Parametric Studies of Viscous and Nonviscous Unsteady Flows, OSR TR No. 55-13, 1955.
41. LIN, C. C., REISSNER, E. and TSIEN, H., On Two-Dimensional Non-Steady Motion of a Slender Body in a Compressible Fluid. *J. Math. Phys.*, **27**, No. 3, 220–231, 1948.
42. MANGLER, K. W., Calculation of the Pressure Distribution Over a Wing at Sonic Speed, RAE Report Aero. No. 2429, 1951.
43. MANGLER, K. W., A Method of Calculating the Short-Period Longitudinal Stability Derivatives of a Wing in Linearized Unsteady Compressible Flow, ARC R & M No. 2924, 1957.
44. Meyer, R. E., On Waves of Finite Amplitude in Ducts. *Quart. J. Mech. Appl. Math.*, **5**, Pt. 3, 257–291, 1952.
45. MILES, J. W., The Oscillation Rectangular Airfoil at Supersonic Speeds. *Quart. J. Appl. Math.*, **9**, No. 1, 47–65, 1951.
46. MILES, J. W., On the Low Aspect Ratio Oscillating Rectangular Wing in Supersonic Flow. *Aeronaut. Quart.*, **4**, Pt. 3, 231–244, 1953.
47. MILES, J. W., Linearization of the Equations of Non-Steady Flow in a Compressible Fluid. *J. Math. Phys.*, **33**, No. 2, 135–143, 1954.
48. MILES, J. W., The Compressible Flow Part on Oscillating Airfoil in a Wind Tunnel. *J. Aero. Sci.*, **23**, No. 7, 671–678, 1956.
49. MILES, J. W., *The Potential Theory of Unsteady Supersonic Flow*, Cambridge University Press, 1959.
50. MOLLÖ-CHRISTENSEN, E. L., An Exploratory Investigation of Unsteady Transonic Flow, MIT Transonic Aircraft Control Project Report 5, 1955.
51. NELSON, H. C. and BERMAN, J. H., Calculations on the Forces and Moments for an Oscillating Wing–Aileron Combination in Two-Dimensional Potential Flow at Sonic Speed, NACA Report 1128, 1953.
52. NELSON, H. C., RAINEY, R. A. and WATKINS, C. E., Lift and Moment Coefficients Expanded to the Seventh Power of Frequency for Oscillating Rectangular Wings in Supersonic Flow and Applied to a Specific Flutter Problem, NACA TN 3076, 1954.
53. ORLIK-RÜCKEMANN, K. and OLSSON, C-O., A Method for the Determination of the Damping-In-Pitch of Semi-Span Models in High-Speed Wind Tunnels, and Some Results for a Triangular Wing, Aeronautical Research Institute of Sweden (FFA), Report 62, 1956.
54. OSWATITSCH, K. and KEUNE, F., Ein Äquivalenzsatz für nichtangestellte Flügel kleiner Spannweite in Schallnaher Strömung. *Z. Flugwiss.*, **2**, 29–46, 1955.
55. RAINEY, A. G., Interpretation and Applicability of Results of Wind-Tunnel Flutter and Control Surface Buzz Investigations. Paper presented to *AGARD Structures and Materials Panel, Copenhagen*, 1958.

130   UNSTEADY TRANSONIC FLOW

56. ROBINSON, A., Aerofoil Theory of a Flat Delta Wing at Supersonic Speeds, RAE Report Aero. No. 2151, 1946.
57. ROTT, N., Oscillating Airfoils at Mach Number 1. *J. Aero. Sci.*, **16**, No. 6 380–381, 1949.
58. ROTT, N., On the Unsteady Motion of a Thin Rectangular Wing in Supersonic Flow. *J. Aero. Sci.*, **18**, No. 11, 775–776, 1951.
59. RUNYAN, H. L. and WOOLSTON, D. S., Method for Calculating the Aerodynamic Loading on an Oscillating Finite Wing in Subsonic and Sonic Flow, NACA Report 1322, 1957.
60. SCHWARZSCHILD, K., Die Beugung und Polarisation des Lichts durch einen Spalt—I. *Math. Ann.*, **55**, 177–247, 1901.
61. SÖHNGEN, H., Die Lösungen der Integralgleichung $g_1(x) = (1/\pi) \int_{-a}^{a} g_2(\xi)\, d\xi/(x - \xi)$ und deren Anwendungen in der Tragflügeltheorie. *Math. Z.*, **45**, 245–264, 1939.
62. SPREITER, J. R., On Alternative Forms for the Basic Equations of Transonic Flow Theory. *J. Aero. Sci.*, **20**, 70–72, 1954.
63. STEWARTSON, K., On the Linearized Potential Theory of Unsteady Supersonic Motion. *Quart. J. Mech. Appl. Math.*, **3**, Pt. 2, 182–199, 1959.
64. STEWARTSON, K., Supersonic Flow Over an Inclined Wing of Zero Aspect Ratio. *Proc. Camb. Phil. Soc.*, **46**, Pt. 2, 307, 1950.
65. TIMMAN, R., VAN DE VOOREN, A. I. and GREIDANUS, J. H., Aerodynamic Coefficients of an Oscillating Airfoil in Two-Dimensional Subsonic Flow. *J. Aero. Sci.*, **18**, No. 12, 797–802, 1951.
66. TOBAK, M., Damping-in-pitch of Low Aspect Ratio Wings at Subsonic and Supersonic Speeds, NACA R & M A52L04a, 1953.
67. TRILLING, L., Oscillating Shock Boundary-Layer Interaction. *J. Aero. Sci.*, **25**, No. 5, 301–304, 1958.
68. WARD, G. N., Supersonic Flow Past Slender Pointed Bodies. *Quart. J. Mech. Appl. Math.*, **2**, Pt. 1, 75–97, 1949.
69. WATKINS, C. E. and BERMAN, J. H., Air Forces and Moments on Triangular and Related Wings with Subsonic Leading Edges Oscillating in Supersonic Potential Flow, NACA Report 1099, 1952.
70. WATKINS, C. E., RUNYAN, H. L. and WOOLSTON, D. S., On the Kernel Function of the Integral Equation Relating the Lift and Downwash Distributions of Oscillating Finite Wings in Subsonic Flow, NACA Report 1234, 1955.
71. WOOLSTON, D. S. and RUNYAN, H. L., Some Considerations on the Air Forces on a Wing Oscillating between Two Walls for Subsonic Compressible Flow. *J. Aero. Sci.*, **22**, No. 1, 41–50, 1955.
72. ZARTARIAN, G. and ASHLEY, H., Forces and Moments on Oscillating Slender Wing–Body Combinations at Supersonic Speed—I. Basic Theory, OSR TN No. 57-386, 1957.

AUTHOR INDEX

SUBJECT INDEX